CONTENTS

Dedication vii
Preface ix
Acknowledgments xi

SECTION 1
Yesterday

Chapter 1 A Brief History of Termites in the United States 3
Chapter 2 Termite Biology Made Easy 11
Chapter 3 Early Development, Tools, and Techniques 17
Chapter 4 Early Chemicals and Their Effects (1930–1980) 23
Chapter 5 Early Media Coverage 27

SECTION 2
Today (1980–2015)

Chapter 6 Treatment Challenges 35
Chapter 7 New Chemicals, Old Techniques 41
Chapter 8 New Chemicals, New Techniques 45
Foam Application 45
Borates 46
Baits 46
Cedar Oils 47
PDS (Pest Control Delivery Systems) 47
Chapter 9 Challenges in Understanding the Label 65

SECTION 3

Tomorrow (2011–????)

Chapter 10 Safety 71
Chapter 11 Toxins vs. Toxemia 75
The Heart 83
The Liver 85
The Gall Bladder 87
The Pancreas 89
Chapter 12 Tomorrow 105
Oh Did I Tell You 114
The Termite Friend Or Foe? 115

References 116
Additional Reading 117
About the Illustrator 119
Picture Gallery 120
About the Author 136

TERMITES YOU HAVE TO WANT TO

YESTERDAY, TODAY, AND TOMORROW

Welcomc to the
World of Termites

JAMES R. MELENDEZ

ISBN: 979-8-88640-653-5 (sc)
ISBN: 979-8-89031-583-0 (hc)
ISBN: 979-8-88640-655-9 (e)

One Galleria Blvd., Suite 1900, Metairie, LA 70001
1-888-421-2397

In the April 2018 issue of the magazine – Business Insider it mentioned the following:

"10 Book's Everyone is Reading Right Now"....

There is another book that can be added to the list.

Termites: Yesterday, Today and Tomorrow by the Author James R. Melendez. Some might wonder why select a book based on termites. Simply the 3 – E'S ...This book is **Educational, Enlightening** and **Enjoyable** to read. The journey of the Author and his friend Izzy, will capture your interest as to why the termite can assist the global earth for enhancing the ecology. While one can only think of a destructive bug, what we learn in this book is indeed fascinating...

My Thanks....to the Author, Mr. Melendez

Howard Swift
Del Norte, Colorado

DEDICATION

As you read the pages of this publication, you will understand why I dedicate this work to our God, Jehovah, who is the giver of "every good gift and every perfect present" (James 1:17, New World Translation of the Holy Scriptures) and no wonder, for Revelation 4:11 (NWTHS) says, "You are worthy, Jehovah our God, to receive the glory and the honor and the power, because you created all things and because of your will they came into existence and were created."

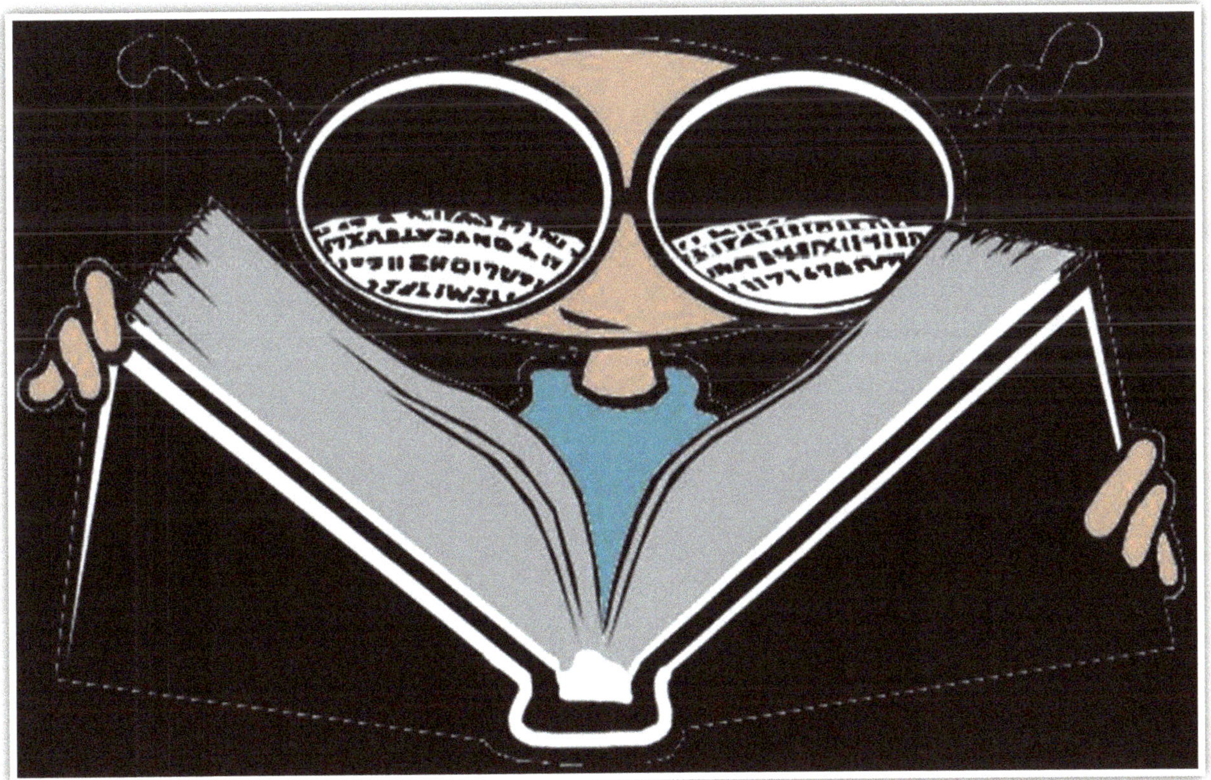

PREFACE

The purpose of this publication is to put into everyday language the challenges that termite technicians have had to endure in treating structures, understanding labels, and dealing with regulatory agencies over the past four decades. While recognizing the importance of academic research and the biological study of termites, for the purpose of this publication *we will not* be putting emphasis on such or on the phraseology of our industry, such as distinguishing between the Eastern Subterranean Termite (*Reticulitermes hageni*), Southeastern Subterranean Termite (*R. virginicus*), Pacific Coast Subterranean Termite (*R. Hesperus*), Arid Land Subterranean Termite (*R. tibialis*), or Formosan Subterranean Termite (*Coptotermes formosanus*). Our emphasis will be on one kind of termite: the subterranean. In our forty years of experience treating termites, we have noticed that, be it a Formo- all have the same foraging habits; the difference is that some are more aggressive than others. While knowledge of the particular termite is important, we feel that the biology of termites in general as well as the structure to be treated, its construction makeup, the type of foundation, and the soil texture *are just as important, if not more so*, to a successful treatment.

We will briefly review and discuss the different kinds of termites and their biology so as to have reference points as we journey through this publication. We will then go back in our time machine to view the history of our industry and how it developed into the monster that it is now.

We are also endeavoring to introduce our industry to the upcoming younger generation and hopefully to encourage them to take an interest in the new technology and advancement of newer chemistry for the betterment of our wonderful planet and for the preservation of the human race worldwide - with a resolved being to be a better place to live. And then we will project into the future using the past and present as our guide.

We are also endeavoring to introduce our industry to the upcoming younger generation and hopefully to encourage them to take an interest in the new technology and advancement of newer chemistry for the betterment of our wonderful planet and for the preservation of the human race worldwide - with a resolved being to be a better place to live. And then we will project into the future using the past and present as our guide.

The language used is not fancy or necessarily scientific, as that tends to diminish the clarity of what is being spoken. Our goal is to use terms understandable to the everyday reader.

ACKNOWLEDGMENTS

In any field of accomplishment, whatever the endeavor, no person can make the statement "I did it all alone," as we are all influenced by our friends, family, background, and education. I'm not any different. I therefore want to take this time to thank, in chronological order, the Arizona Pest Control Board, which helped me get started in the early seventies; the Arizona Structural Pest Control Commission, which helped me understand the legal aspects of our industry; and the Utah Department of Agriculture and Food, who has one of the best support systems for our industry that doesn't violate our dignity.

One of my early experiences that had a lifelong influence on my career was meeting Dr. Austin M. Frishman, the grandfather of pest control and a pillar of our industry. He told me, "*You have to want to,*" and that was forty-five years ago. That was my goal then and still is now, as it was in agreement with my moral teaching as found in the Bible. Thank you, Austin and your wife (see photo in the picture gallery).

This acknowledgment would not be complete if I didn't mention my bride of forty-seven years (Nancy), who supports me every step of the way. Without her, nothing would have been possible. She knows more than I, but she will never attest to it. She is my walking field manual and insect identification guide. Thank you, sweetheart.

One of the best accomplishments parents can have is to see their children follow in their footsteps, and it fills my heart with joy when my daughter Nicole is at my side at every seminar, as she has been since she was in diapers.

We also thank Target Chemicals, FMC, Dupont, BASF, and all in the chemical industry who deadly insect- and vector-borne diseases. All the above is only possible because of the National Pest Management Association, who takes the lead in protecting our planet.

Thank you, all.

* * *

All scripture quotations are from the *New World Translation of the Holy Scriptures* (NWTHS), revised 2013, published by Watchtower Bible & Tract Society of New York, Inc.

All dictionary references are from the 1994 edition of the Merriam-Webster Dictionary.

SECTION 1

YESTERDAY

T

CHAPTER 1

A BRIEF HISTORY OF TERMITES IN THE UNITED STATES

As mentioned in the preface of this book, the information is based on over fifty-three years of my hands-on experience in the trenches of the field of pest control. As I emerged into the pest control industry in the early seventies, I was fascinated by the biology of termites and their relentless determination to always stay one step ahead of me. I can still hear them laughing, but their end is near.

For those companies and technicians who are interested in fulfilling their moral obligation to their client, their company, and to themselves, the business of managing termites, or termite control, has always been a business of technology. I entered the scene in the seventies, but my studies took me back to the forties, where I found information written by experts like Frank Museum of Natural History. In his article "The Truth about Termites," he said that exterminating racketeers had labeled them "public enemy number one." These white ants (which aren't ants at all) are seldom destructive in civilized communities and are definitely constructive in nature. Incidentally, some species can't digest wood any better than we can.

Even though they are sometimes called "white ants," they are neither ants nor even closely related to ants. In addition to a large number of structural differences between ants and termites, there is an important difference in their life history: ants hatch from eggs as worm-like creatures and go through a pupal stage before becoming adults. Termites, upon hatching from eggs, look much like small adults (except that no newly hatched insect ever has wings)—they do not go through a definite, apparently lifeless pupal stage in their life history. In other words, ants undergo complete metamorphosis* while termites undergo incomplete metamorphosis**. I suppose that there is no real objection to calling them "white ants," provided we remember that they are not ants and that they are not white. We will call them termites in this book.

All termites belong to the order Isoptera, and all Isoptera are termites. If we say that they are related to cockroaches we shall not be nearly so wrong as if we relate them to ants. In fact, we might go so far as to say that they are social roaches.

Can you visualize what was going through my head as I embarked on this new mode of thinking when I learned more about these destructive insects? I said to myself, *This calls for war on the little critters.*

Further study of the proper procedures guided me to a book called *Urban Entomology* by Walter Ebeling, which included a chapter on wood-destroying insects and fungi. Please bear in mind that this was before the days of the Internet and Google, when libraries ruled. In his book Ebeling gives good guidelines on procedures, which I found very useful, including information such as the following:

> Termites feed on wood, and throughout a large area of the world they are the most destructive insects to wood structures. This is a measure of the importance of their role in nature—breaking down and returning to the soil and atmosphere the enormous tonnage of dead and fallen trees and other celluloid material that are continuously accumulating on the earth's surface. They are important pests of agricultural crops, forest nursery seedlings, range grasses, and they also damage an enormous amount of stored food, and household furniture and commodities, including even most plastics (Snyder 1935, 1955b; Harris 1961; Ebeling 1968). Damage from termites, plus the cost of controlling them, probably amounts to a half billion dollars per year in the United States alone (Ebeling 1968). Our knowledge of the history of South America would probably be much more complete if it had not been for termites, for they are said to have eaten most of South America's centuries-old books (Howse 1970).

* Complete Change

** No change just glow up from small to big

Termites are found in tropical, subtropical, and most temperate climatic zones. They are increasing their range and density northward, favoring the accelerated urbanization—the "population explosion," particularly in areas where central heating in buildings affords them a favorable environment for the establishment of colonies.

It[*] also gives procedures for rodding, sub-slab injection, interior wall gassing, and inspection procedures. (See chapter 3: Early Development, Tool, and Techniques.) At that time, Chlordane was the product most used, and, I must say, it did the job and did it well. Some houses are still being protected by it, and it has been in the ground for over thirty years. 80 percent of the procedures used for the control of termites was based on the longevity of the compound organochloride better

* That is period the book-Urban Entomology by Walter Ebling

known as chloradane. It is interesting to note that prior to chemical treatments, homeowners would just remove the damaged wood and replace it with new lumber. Also, prior to chemicals, construction methods were quite different. For example, many older homes, especially along the east coast of the United States, have a porch along the front entrance. Why? Because they had a crawl space! Builders knew from experience to keep as much of the living floor space off the ground to prevent termite infestations as well as to provide access to plumbing, sewer, electricity, and the like. In my opinion, pear construction is still the best when it comes to termite protection and accessibility of utilities.

In summary, the history of termites continues until this day as it did when man first noted termites' aggressive and vigorous appetite. Frank E. Lutz said it well:

> It is interesting to note that termites are among those insects that Nature uses to clear away stumps. If man, when building houses, puts lumber (which is nothing but dead wood) in or on the ground, the termites, quite naturally, are apt to start clearing away that dead wood. Man's wooden houses, and even man himself, are things of the last few days or so, compared with the geologic history of termites. If man keeps the wood of his houses in this region sufficiently high above the ground our termites will not bother him.

1912 proved to be the early signs of goodbye for us termites with your introduction of the spray can. we felt like we were getting our pink slip, You just don't understand us.

1926 In 1926 we could see the hand writting on the wall when you learned about wood to earth cantact. we realy worked hard at keeping that a secret.

1922 In 1922 you incorporated a building code, but we didn't worry about that because you humans are not to good at following instructions.

1931 Ok Jim! Says Izzy and shaking his antenna's. stop right there, Startled, Jim answers: what happened now Izzy? Well- let me tell you! While we're trying to keep your forest clean so you could go camping and building fungus farms to balance the decaying process and not to mentions the ozone we put into the atmosphere for your benefit, and what do you do? What are you talking about Izzy! OK, what did you do in 1931? Jim answered, I don't Know. Izzy pacing the wood beam back and forth and strumming his antennas back and forth, says, well-for your information that's when you humans started to use creosote and all our hard work on the railroad ties came to an expeditious end, and how do you think we felt?

1932 That was the year we starded our anxiety attack.
Why? says JIm.. Why you ask? Because thats the year you starded your frontal attack. Frontal attack says Jim. Yes, Izzy nervously answers. thats the year you starded soil treatment.

1936 And that's not all: now to add insult to injury- you go and introduce arsenic to continue you war against us. When are you Gona get it? Get what? Jim replies. Izzy baffled by that reply answers. we are on your side.

1949 In 1949 you introduced cloradane, and we had to endure from 1950 until 1988 when cloradane was removed from your termite aresenal. we were so happy to see the organophosphtes. What a sigh of relief.

1994 Around that time we starded to see the development of baits, but they didn' scare us, what did scare us were the Borat wood treatments, and what through us into a panic and terrifed us was the development of the Delivery System. Not only for us but for all of our athropod cousins.

It makes us happy you humans don't like to apply yourself to doing a commendable job as we see it, because if you did more of you would be installing delivery systems.

CHAPTER 2

TERMITE BIOLOGY MADE EASY

Hello, my name is Izzy. Well, they call me Izzy, but my real name is *Isoptera Termes* (I'm an equal wings termite), and I would like to take a few minutes to introduce myself. As I mentioned, my name is Izzy, and I'm a termite that belongs to the order Isoptera.

I have four wings of equal size, and, believe me, it makes it really difficult to fly. In fact, it's impossible to control my flight pattern. That's why people say we don't fly; we just flutter. (Boy, do I hate that!) Because I just flutter, I depend on the wind to get to where I'm going and to start a new family—you humans call them *colonies*. Once I arrive at my new location, I find a mate and start rearing my family. I call my mate my queen. She calls me her king. She really likes that! When we mate, it's a lifelong arrangement. We remain together indefinitely and continue to mate throughout our lives.

We are very social insects, and you can find us throughout the world. Some say we only live in warm climates, but don't believe it. You can find my family even in cold climates like the Rocky Mountains of the United States and also in places like Monticello, Utah, where the elevation is seven thousand feet above sea level and winter temperatures dip below freezing.

In my family there are many different kinds. You may not believe it, but over two thousand termite species have been cataloged—approximately forty species in the United States alone—so you can understand why we are of economic importance.

Consider this: we cause over five billion dollars of damage to US homes each year. Just one, small colony of approximately sixty thousand of us can eat one linear foot of a two-by-four in five months. In some regions of the United States my cousins—the Formosan subterranean—have colonies that typically number in the millions and can travel over one hundred feet to a food source. We subterranean termites cause over 90 percent of termite damage in the United States.

We begin our flights (swarming) in January in the southern states and in May or June in the north. Some of us can chew through lead, asphalt, plaster, or mortar to find wood. We are serious about the responsibility we were assigned when we were *made* on the sixth day of Creation.

If you were to put all us termites together and weigh us, we would outweigh humans on Earth. My queen can live fifteen to twenty-five years and can lay an egg every fifteen seconds.

I want you to know that our children, the worker termites, feed on both sound and decaying wood or other plant materials, such as humus, grass, fungi, and paper. The cellulose in these foods is digested with the help of protozoans* that live in our intestines. Members of the colony re-infect each other with protozoans as they exchange feeding substances. Without the protozoans, we can't survive. You might liken it to the human digestive system. If humans can't digest food, they become sick, and the result could prove to be fatal.

We have three body sections: the head, the thorax, and the abdomen. We have a broad "waist" between the thorax and the abdomen. We have a single pair of straight antennae with bead-like segments. When eyes are present, they are either compound or simple, but that doesn't make a difference because we are blind. When we travel through our mud tubes, we navigate by the use of pheromones, so it doesn't matter to us that we are blind. Moisture is vital for our survival. We obtain our moisture primarily from the soil and thus must maintain contact with it to survive. We move underground through an extensive gallery system, through shelter tubes or mud tubes (as I mentioned previously), and invade wood separated from the soil. Our shelter tubes are constructed by our workers from particles of soil or wood held together with fecal material.

* one cell microscopic animal

When there is an above-ground source of moisture, we can remain in the wood without ground contact. When there isn't enough moisture, we have to return to our nest because, you see, we can't dry out. If we did, we would be history. (See Photo Gallery.)

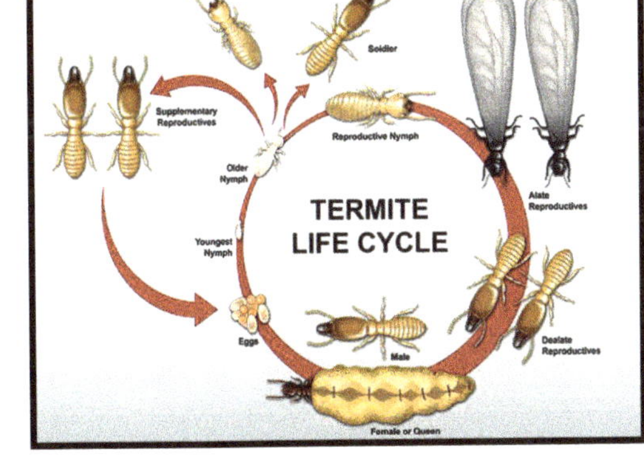

Oh boy! Guess whom I forgot and overlooked? I'm ashamed to face you, but I'll try to make it up to you. Do you remember that I told you we were social insects? Obviously, what I mean is that we are organized. We live in colonies within the ground and have specialized castes to perform specific colony functions. Our colonies have three primary castes: worker, soldier, and reproducer. The soldier is the one I forgot.

Soldiers have elongated yellowish heads with large jaws and are about the same size as the adult worker—a quarter inch. There are fewer soldiers than workers in a colony, and they rely on the workers to feed them. Whenever the colony is invaded or a hole is made in a tube or piece of infested wood, the soldiers will use their jaws to defend the breach. What would we do without them? Because soldiers play such a vital role, let me explain their social behavior in more detail.

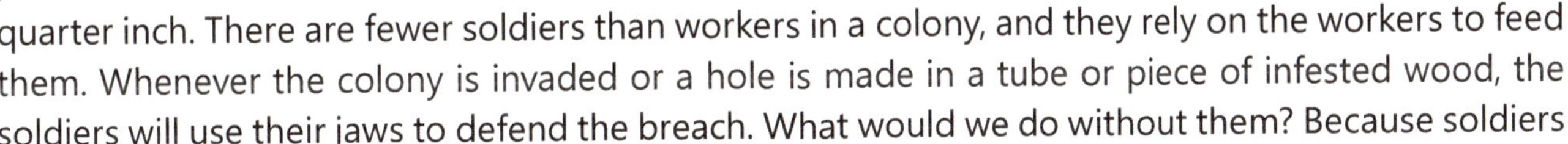

With social insects, communication is necessary to maintain efficient integration and division of labor. The most basic means of communication we use are via chemicals (pheromones), which I mentioned earlier. In fact, each colony develops its own characteristic odor. Any intruder, be it a termite from another colony, an ant, or any other natural enemy, is instantly recognized as foreign when it enters the colony. What happens then? An alarm pheromone is secreted by the colony that triggers the soldier termite to attack and kill the intruder.

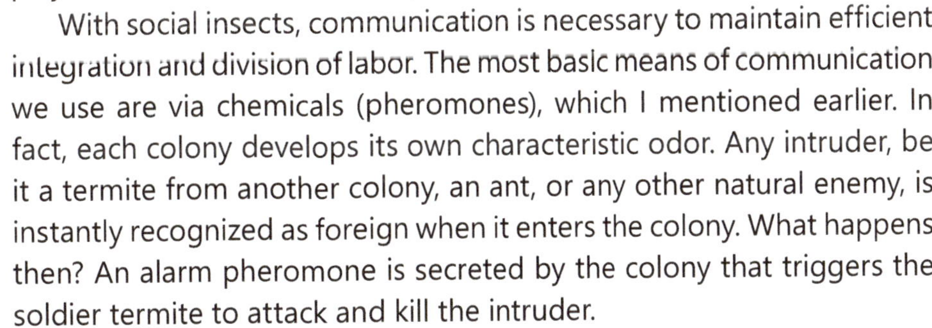

The intruder is then walled off from the colony with fecal matter. If a hole in the termite workings occurs, the workers immediately patch it. Do you now understand why I felt so bad about forgetting the soldier termite? What would we termites do without them? Although they are few in number, they are critical for the survival of our colonies. It's well that ends well! However, it hasn't ended. Next I want to talk to you about the early development of tool and techniques humans have used to protect their buildings from termites.

Izzy's playground

Did you say termite tube

Izzy's Workers

CHAPTER 3

EARLY DEVELOPMENT, TOOLS, AND TECHNIQUES

Ha, ha, ha—I can't believe it! Haa, haa—would you believe it? I'm sorry for laughing, but I can't help it. They want me to talk about early development, tools, and techniques of dealing with termites when we should be talking about old-fashioned ethics and standards. You might think that because we live in the ground and are lowly termites, we're devoid of human ethics and standards, but remember we were around long before you and have been observing your behaviors. Believe me, you have nothing to be proud of!

I can remember when Egyptians put cedar sawdust and cedar chips on the ground of their tombs and houses to keep us, our cousins, and our friends (other insects) out. That practice prevailed for thousands of years.

Now let's put our time machine into high speed, forward! Find the nineteenth and twentieth century on the right side of your screen and press Enter. Make sure your safety harness and belt are secured tightly. Press Enter and hold on. Upon arrival, the machine will turn off automatically. As you exit the machine you will immediately notice that although humans have made major advances in technology in all fields—not only in termite technology—when it comes to ethics and moral standards, humankind has fallen behind, thus making it easier for us to sidetrack you and outsmart you (ha, ha!), not to mention all the daunting challenges you have to face after the smoke clears. We just stand back and laugh. That's why I was laughing at the beginning of this discussion.

In the forties, fifties, sixties, and the first half of the seventies, whenever we saw a termite exterminator approach one of our high-value meals (often someone's house), we knew we had to move fast and that most of the time we wouldn't make it. It wasn't the chemicals we were afraid of—it was the desire of the operator to do a good job that frightened us. We knew the operators would find it unacceptable if they violated their God-given conscience, moral ethics, and standards.

Some of tools in the arsenals of last century's exterminators, such as dusters, compression tanks, sprayers, power sprayers, compression needles, and drills, are still being used today. However, you have to remember, you have to want to! Sometimes I think we are playing hide and seek, and I think we are winning. Ha, ha.

Some of the early use of the equipment mentioned is much the same today, as can be seen from an examination of Ebeling's text:

Application Procedure* and Dosage**

> Termite operators generally apply insecticide emulsions with power sprayers but high pressure is not required and it is possible to apply the emulsions with a sprinkling can. To treat the ground where a concrete slab is to be poured, including attached porches, apply 1 gal. (about 4L) of the diluted insecticide emulsion to 10 sq. ft. (about 1 sq. m.) as an overall treatment. Overall treatments applied to washed and ungraded gravel fills, or fills of an absorbent material (e.g. cinders), should be increased by 0.5 gal. per 10 sq. ft. (2L per 1 sq. m). Apply 2 gal. per 5 linear ft. (7.5L per 1.5 m) to critical areas, such as along the in- side of foundation walls and around utility pipes, entrances, and interior partition foundation walls, also use 2 gal. per 5 linear ft. along the outside of the foundation.

See pictures.

As seen from the above, the techniques and tools are very similar. So why are there so many controversies when it comes to termites compared to ants or cockroaches? Because when it comes to termites … you have to want to!

> From the crawl hole that provides access to the attic, the dust is usually applied with an electric blower that generates an airstream of high velocity or with a water-type fire

* not necessarily today's procedure

** dosages vary according to the label direction

extinguisher, commercially available for the application of insecticide dusts (chapter 3). Fire extinguishers are filled with dust to within a few inches of the top and then pressurized to 100 psi. Because they are equipped with a Schraeder valve, they may be pressurized at a service station or by other sources of air or gas, such as small electric air compressors or tanks of nitrogen, provided the latter are equipped with the appropriate type of outlet. They can discharge a high volume of dust at high velocity and, like the electric blowers, can dust the average attic from the crawl hole. The unusually light weight of silica aerogel results in an even distribution of dust throughout the attic and into its extremities.

In addition to attic dusting, the protective film should be extended to include wall voids and other enclosed spaces. This is best done during construction when holes can be made in the plaster lath about 4 ft. (1.2 m) above the floor and between every two studs (Ebeling and Wagner, 1964). The holes may be made with a sharp pick or a half-in. (13-mm) drill. Three grams of silica aerogel are blown into each hole. If the plaster lath is already perforated there is no need to drill holes. Even with perforated lath, which is currently rarely used, the bulk of the dust settles on the interior of the void or falls to the floor plate.

If insulation is present we recommend that the dust be discharged downward, between the plaster lath and the paper that encloses the insulation, to be deposited principally on the floor plate. When "dry-wall" construction is used over conventional wood-stud framing the sheets of "sheetrock" must be installed horizontally which provides a horizontal juncture of sheets 4 ft. (1.2m) above the floor. Holes through which treatments are made are drilled at this juncture and between all studs.

Termite galleries are located by probing suspected timbers with a screwdriver, ice pick or other sharp instrument. Quarter-inch (7 mm) holes are then drilled into the infested wood members at about 1-ft (30-cm intervals to provide access to the galleries. An insecticidal dust such as "Kali- dust" (50%calcium arsenate) may be blown into the holes by means of a "Kaligun") or a liquid fumigant may be injected. An ounce (28cc) of dust is enough for 15 to 30 holes; too much dust may plug the galleries and decrease the effectiveness of the treatment. If you understand the ABOVE COMMENT YOU ARE a GENIUS.

Block Construction

We Love It!

Come and get me

Did you say hide and seek

NOTES

CHAPTER 4

EARLY CHEMICALS AND THEIR EFFECTS (1930–1980)

IZZY. Hi! Now how do I start this and make it easy for you humans to understand? Well, do you remember me talking about the ark—that is, Noah's ark—and remember I mentioned pitch and it being a hydrocarbon?

JIM. No, I don't remember.

IZZY. Oh, that's right. That's not until the end of the chapter. Well, let me explain that. Better yet, why don't I let the wise geek explain it. You see, I'm not too smart when it comes to human science, and if I spoke in termite terms you wouldn't understand it.

WISE GEEK. Organic molecules that consist of a chlorine atom with hydrogen and carbon atoms are called chlorinated hydrocarbons. These have several uses, from cookware to solvents. An example of a chlorinated hydrocarbon made in a laboratory environment is vinyl chloride. However, these chlorinated hydrocarbons can put humans and the environment at risk. They can cause health issues and other problems if exposed to the environment. In some cases, some chlorinated hydrocarbon pesticides like DDT appeared safe to use, but scientists later discovered the pesticides compromised bird populations: it affected eggshells, which made it hard for eggs to hatch.

IZZY. You may see now why I didn't want to explain it!

Jim, *tilting his head, and squinting.* I'm not sure I put on your thinking cap.

The book *Insight on the Scriptures* states: "Pitch—the sticky liquid form of bitumen, a dark-colored hydrocarbon similar to what is generally called tar"

Izzy. Did you notice that pitch, or tar, if you want to call it that, is a hydrocarbon?

Jim, *stretching his neck and head to the side.* Yes?

Izzy, *reaching out to him.* Are you starting to get it? These chemicals are not new. That is why we termites were not able to penetrate the ark, even though it sat on the ground for decades before it started to float. Boy, were we angry! What a morsel, and we couldn't touch it; and that's why chlorinated hydrocarbons were so popular until the eighties when dealing with termites. The organophosphates then began to rule for a while. *What a relief for us termites!*

The organophosphates then began to rule. What a relief for us termites!

NOTES

I want to remember or forget.

CHAPTER 5

EARLY MEDIA COVERAGE

We termites are perplexed!

With all the damage we cause (and we cause billions of dollars of damage a year just in the United States) and the media being completely absentminded, acting as if they don't care (unless of course it's in the White House, some other government building, or in the home of celebrities like Hansel and Gretel), we can't figure it out—believe me, we've tried! We've damaged buildings in every state of the union, not to mention the defacement we caused to your personal property: art collection, books, and the like. I think you already know (and if you don't know, you will not find it in the media unless you are Dr. Watson!). Read on, and maybe you'll be able to understand why we are so perplexed.

As temperatures start to rise in springtime, we termites begin our busy season just the same as you humans start to prepare your gardens, but our gardens are your houses and whatever we can find that accommodates us well. The difference is, though, that you usually can't see us.

Now, unless you know what to look for, you probably don't even know what is happening. We even have you thinking that we are flying ants so you believe that all is well! So sorry, so sorry! For example, in the United States, we damage 687,000 homes each year. In Miami, Florida, alone— where we do the most damage, we love it!—the amount of wood our family consumes annually would build thirty-one homes. Now, did you hear that in the evening news? I don't think so! However, it was mentioned in *USA Today* by Sam Ward—not on the front page, but stuck in the back, in the "Help Wanted" section.

Are you getting it? Well, if that's not enough, read this:

Termites gnaw FDR drive pilings: "Gribbles" have major case of gobbles

Some folks call them "gribbles," but don't let the cutesy name fool you. Marine borers were banished from the city's waterways in the 1940s by pollution. However, cleaner discharge from better sewage treatment plants has allowed them to return in full force, leaving fallen Harbor—the *limnoria*, a flea-size shrimp-like organism, is more plentiful in the East River. It is believed to be the main culprit attacking wooden pilings supporting the FDR drive and nearby piers. It attacks the outside of a wooden piling and burrows in, reducing strong pilings to flimsy wooden hourglasses. "We can't tell if they're eating the wood for nourishment or just making a home", said Dr. Paul Boyle, Deputy Director of the New York Aquarium in Coney Island, Brooklyn. Shipworms, which can be as long as two feet, are dominant in the Hudson River. The shipworm bores into a pile and stays there, eating its way through the wood on the inside. They are believed to be playing second fiddle to the *limnoria* on the FDR. Marine borers have their place in history: in 1503 they sank two of Christopher Columbus's ships. To avoid such mishaps many piles around the city were coated with a protective tar, but that tar has worn away over the decades, just as the ravenous creatures returned. Recent studies have shown their population growing by up to 8% a year in New York Harbor. Their dirty work caused the closing of the Hudson River's "Penny Park" near Tribeca for nearly a year as pitched six people into the water. (Rutenberg 1999)

GRAPHIC: THE BUG THAT ATE THE FDR DRIVE Aquatic wood borers the size of a flea are eating away at 45% of the pilings supporting the FDR Drive. The shrimp-like *limnoria* (right) attacks the outside of the pilings, leaving behind flimsy wood hourglasses.

The city is spending $6million to find a way to stop the pesky creatures. Cross section of FDR Drive and the East River Source: New York State Department of Transportation.

Now do you get it? When we do get coverage, a good portion of the time it's not completely correct; no wonder you humans are so confused about us—and believe me, you are confused. Did you notice that the word *termite* was not even mentioned in the article but the word *limnoria* and the *shrimpworm* were? Now look again at the title of the article: "'Termites' Gnaw FDR Drive 'Gribbles' have a major case of Gobbles."

1978: Formosan Termites Invade New Orleans

The Formosan termite hitched a ride to New Orleans and other southern ports after World War II, an unwelcome stowaway in cargoes of returning war equipment. *Coptotermes formosanus* is a voracious species, devouring wood nine times faster than native termites. For twenty years they burrowed into the soil, establishing multimillion-strong colonies. In the late 1970s they burst out of their isolated footholds and began to eat their way through south Louisiana and beyond.

Nearby the port where they arrived in this country, the French Quarter's common walls and wood framing made a perfect *Formosan* habitat. The termites caused massive damage to the Quarter until the arrival of new baiting systems in the 1990s. At a sign of activity, slow-acting poison was loaded into a subterranean bait trap. The termites carried the bait back to the nest, killing the colony. The baits have slowed the damage Formosan Termites cause. Some termites in a colony grow wings and swarm out of their nest on humid nights in late April and May. They are attracted to any light.

Formosan nests in a building need not be in contact with the ground, making the traditional trenching and drenching with insecticide barriers less effective. The insects can bypass traditional barriers and continue to attack the same structure. Their colonies can extend 300 feet underground.

Unlike an infestation of native termites, which can be killed by cutting off their source of water or path to the ground, often the only way to get rid of a Formosan infestation is to tent the entire house and gas it. However, the treatment has no effect on the giant colony that may be lurking beneath the ground.

If we had our way we would have a special section in every newspaper on termite damage alone. We could call it "Izzy's Termite Corner!"

As you can see, for all the work we do, we really don't get the attention we deserve until it's too late.

Let us get back into our time machine and find the 1980s. Now find the Slow button, and then press Enter. You will not need your harness because we are just moving around the corner a few years. Wait until the time machine comes to a complete stop. As you dismount look for the path that says "Treatment Challenges," and just a few steps to the right you'll find some buttons. Press the one that says "The Label is the Law."

SECTION 2

TODAY (1980–2015)

CHAPTER 6

TREATMENT CHALLENGES

Well, just think of it! We are so vexed that you humans have to write laws to keep us in check. I say "check" because you can't outnumber us or outsmart us! Our biggest fear is not the chemicals or what the label states regarding proper application; what makes us tremble in our tubes is the technicians who really want to (as we mentioned, you have to want to).

In some older homes it's just impossible to penetrate the soil. When faced with that predicament you must document it in writing, if possible with pictures in your paperwork, and notify your client. It doesn't necessarily mean you are not doing a good job. The fact that you are documenting it is proof that you *want* to. You also want to keep in mind that if the chemical can't go down, we termites more than likely cannot come up. The judge will understand!

If you violate any part of the label, it's like telling a police officer you weren't speeding when you were. Notice what *PCT* magazine says (Pest Control Technology) on the March 1990 issue's front cover:

Making sense of the termiticide labels

Unclear or inconsistent label language can make treating a structure difficult. What is the answer? See chapter 9.

We termites love it. Why? Because if you are confused before you begin, how do you know whether you are doing a good job? So, how can you want to if you don't know how to? The answer: *you have to want to*!

As mentioned earlier it is imperative to know and understand proper construction methods and protocols. These codes vary from state to state and county to county as well as from construction company to construction company, since people's values differ. Another very important factor is the time of year of the construction project. For example, slab poured during hot or summer temperatures is more apt to crack, thus allowing us termites to enter; we call it "the open-door policy." We love it!

That it takes longer for the concrete to dry and it is usually protected by using blankets or straw. The rule of thumb is the longer it takes concrete to dry, the harder it becomes. Cracks can also develop if concrete mixture is not correct or compaction was not executed correctly. Sometimes pipe and sewer trenches are not compacted prior to slab compaction, thus allowing for crack development. That occurs because as both hot and cold water passes through pipes and sewers, condensation forms along the exterior of the pipes, creating water droplets; the results: you now have compaction all along the pipe trenches!

Thus the soil around the pipe settles, leaving a gap between the top of the soil and the bottom of the slab; that is where we termites love building our tubes! Now all we have to do is to wait for a heavy truck to go by, like a garbage truck or eighteen-wheeler, and then, poof, the crack is born. The open door policy! Charge! The last one there is a rotten egg! Don't forget Mabel!

In the living room when you say: Why are they here when there isn't any water or moisture here? I can understand the bathroom, but why here? What happens is this: you have bathrooms at both ends of your house or the kitchen or laundry room at the other end, all connected under your slab to a septic system or to a main sewer line emptying into the city sewer-line. Again, if the trenches are not compacted properly, you will be inviting us for dinner, to have "a covered dish" (a covered wall). Do you understand now why this chapter is entitled "Treatment Challenges"?

NOTES

I want to remember or forget.

Another treatment challenge is dealing with eight-inch-block construction walls or worse when eight-inch blocks are used as footers with a wall plate on top. For us termites that is our favorite, as we don't have to wait for cracks to develop because you are building with cracks: cracks between blocks and cracks in the blocks—that is, the hole in the block. No matter what you may try to use to fill those voids, or how you put those blocks together, you will never be able to seal them off 100 percent! All we need in order to get in is a crack the size of one of your hairs; that is why we love block construction.

Worse construction – Blocks

Damage we cause, but it will be difficult for you to discover our freeway system (tubes). To treat the structure properly you will have to drill into every block joint, and then, maybe, you might get us. Not only do you have to want to, you almost have to be crazy to want to *want to*. If you understand that, then you are on the way to understanding us termites.

Another one of our favorite construction materials is foam block construction. That is where you pour concrete in forms made of heavy-duty Styrofoam, connected with metal clips. They say it is good for insulation because, you see, you don't remove the Styrofoam but leave all the forms in place. They forgot to tell you this is good for termites too, the reason being that we can find our way between the concrete and Styrofoam because of the moisture created by both. Now it's time to play "Hide and Seek" and "Hope against Hope" (to hope without any basis for expecting fulfillment).

We termites want you to know that we take a lot of pride in the work we were created to do. We are not trying to cause you any harm! We are just trying to be any harm, maybe in a symbiotic relationship (see chapter 6, Chart One, "What you will see us do as we treat your home for termites").

CHAPTER 7

NEW CHEMICALS, OLD TECHNIQUES

"No one puts new wine into old wineskins. If he does, the new wine will burst the wineskins and it will be spilled out and the wineskins will be ruined. But new wine must be put into new wineskins" (Luke 5:37–8, NWTHS). Thus said the greatest man who ever worked on this planet, Jesus Christ. Now I (Izzy) was there when Jesus spoke those words, and he knew what he was talking about. That illustrates well what happened in the mid-eighties when CHCs (chlorinated hydrocarbons) were fired and were replaced by organophosphates and later by pyrethroids.

When CHCs exited the back door—and quit abruptly at that when you think about it—it left you humans with few choices. The techniques that should have exited with them stayed behind and lingered, not that the techniques were wrong in themselves, as they worked well with the CHCs. All that was needed was a re-evaluation in light of the difference in chemical longevity of CHCs compared to the chemicals that replaced them.

The scenario wasn't negative or intentional. All that was needed was an adjustment in the label instructions recommending a re-treatment every four years. An explanation to the client on the longevity of the new compounds would have been satisfactory and acceptable if the client knew beforehand, but it was not in the exterminators' control. Instead they tried using the same paperwork and guarantees they used in the days of CHCs even though re-application is now a norm in the pest control industry. We termites were in our glory, not only because of the damage we caused to structures but because of the havoc we caused in the law courts. Some cases were absolutely hilarious. Companies were treating homes in an identical manner to the fifties because the treatment procedures specified on labels hadn't changed. Now let me ask you: What do you think would have happened if your highway systems hadn't changed from the thirties and forties to keep up with your automobiles, with all their power and speed? Today there would be the road engineers who can see the future and who are trying to make peaceful changes. There really isn't much more to say about new chemicals and old techniques except that we termites really like you when you don't really want to. You can't blame the chemicals for a scruffy job. The chemicals only go where you put them or where you really want to put them, because *you really have to want to.*

In the chapters that follow, Jim (with help from Izzy) will explain in more detail the birth of the new application treatments and how Izzy and his extended family feel about them.

CHAPTER 8

NEW CHEMICALS, NEW TECHNIQUES

Those definitions fit our industry well. The technical details are not the tools or how to use them but what it is that you are trying to accomplish. In our case, it's protecting the structure while protecting humans and our environment.

It is good to note that the protection of human health and the environment has always been a concern of our industry, but with the exit of the CHCs it has taken on a higher level of concern, and rightly so in view of the last part of Revelation 11:18 (NWTHS), which states: "to bring to ruin those ruining the earth." The manufacturers, the pest control community, and the media are now playing found, and it is also one of the main concerns of the National Pest Management Association. That endowment has given birth to a new era. This new era and its affiliation has influenced technology to develop foam application, borates, baits (they are not really baits but monitoring stations), cedar oils, and the improved sub-slab injection PDS (pest control delivery system). We will discuss these one at a time.

Foam Application

In the process of injection, foam is introduced into the chemical as it is being applied in order to maintain an even distribution of the compound in the void being treated. Izzy informs me that some companies would use soap to accomplish the same results before the introduction of foam machines. If anybody would know, Izzy would! Thank you, Izzy.

Borates

With the introduction of the borate chemistry, our industry has taken a huge turn for the better in regaining the trust of the public. Here is what Wikipedia says about its chemistry:

> Disodium octaborate tetrahydrate, an alkaline salt, is produced in two forms. One is a clean, liquid concentrate which can be mistaken for corn syrup if repackaged and not stible or explosive and has low acute oral and dermal toxicity. This salt, which is commonly confused with boric acid, is used as an insecticide [1] and fungicide, and is commonly sold as an insecticide in liquid or powder form. It is also effective against fungi and algae. It has an infinite shelf life and is not affected by temperature. "Bora-Care" is the liquid form suitable for use in a sprayer. The powdered brands are "Tim-Bor", "Borathor" or "Termite Prufe" as sold in the United States. This chemical is also a flame-retardant [2].
>
> In liquid form it is commonly diluted and sprayed on wood surfaces to kill termites, carpenter ants, fungi and algae. This alkaline salt is not to be confused with Boric Acid (an acidic chemical) or the laundry detergent additive used for stains.
>
> In common use as a termite control, or for termite prevention, the liquid concentrate is used at a ratio of 1:1 with water. The advantages of this chemical over conventional pest control treatment is that it is non-carcinogenic and has a low toxicity to humans and pets. It is also odorless and proper application lasts for the lifetime of the wood. Repeat treatment is not necessary. It is best used during new construction. However, it is more commonly used after the fact on homes originally treated with chemicals that have become inactive. It was shown to significantly reduce dust.

Baits

Another positive development has been the introduction of what we call bait stations, which are not bait stations at all because they don't bait termites with the intent of killing them. When we think of bait, what comes to mind is an object or maybe food that will lure your prey into your trap.

These stations don't function as traps or lures, but they do a great job of monitoring termite activity when placed around a structure. Izzy informs me that he and the other members of his family are not attracted to them and don't look for them, but if they run into one they will stop for lunch. This could be a good indication of a termite colony nearby. When bait stations indicate termite activity, it's time for an evaluation to determine whether or not to perform a treatment.

Cedar Oils

Cedar oils are the new guys on the block. We say new, but they are not new at all. Remember Izzy said the Egyptians used cedar chips on the ground to keep insects out of their tombs, including termites? Well, it still works; Izzy says, "We hated it then, and we hate it now!"

PDS (Pest Control Delivery Systems)

Wise King Solomon said in Proverbs 13:16 (NWTHS), "The shrewd person acts with knowledge, but a fool exposes his own foolishness." Similarly, Richard P. Feynman said, "The only person who is educated is the one who has learned how to learn and change, thus said not be fooled," (Feynman). When you zero in on an infestation, or you think you did, Izzy and his dinner companion are very happy when they see you come home, pull back the carpet, drill again, and treat again. Why are they happy? Well, they know you just didn't want to. Now if you had installed a PDS, your homecoming would have been more pleasant. Why? Because all you would have needed to do was recharge the system.

The PDS consists of permanently installed piping or flexible tubing to be used to deliver the product to critical inaccessible areas under the slab, such as concrete expansion joints, cracks, and plumbing utility services penetrating the slab.

Another use of the PDS is for crawl spaces. A label might read as follows: "product may also be applied through under-structure insecticidal delivery systems such as piping or flexible tubing mounted under the structure." A word of caution: Izzy informs me once again that "if the system is not closed-looped, the hydrodynamics will not equate, leaving void spots, and we termites have a great time and savor every morsel of.

Post Construction

- **PDS Installation**

Lay out drill points according to label

Drill points-Notice cold joint crack

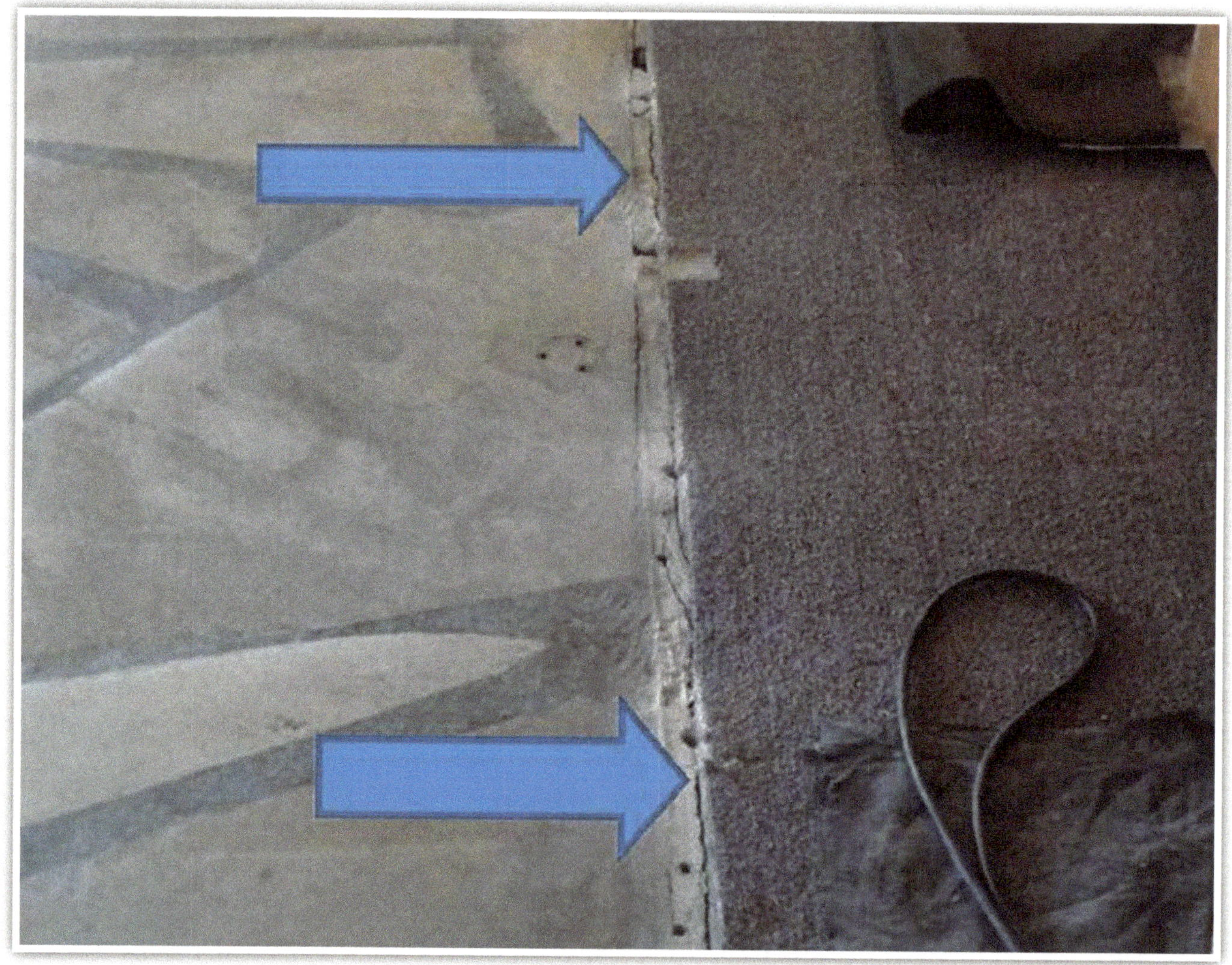

Completed drill points and ready for PDS Pre-fabrication

Work clean

Assemble ¼ inch PDS line to fit drill points

Install as you go along

After completion- remove

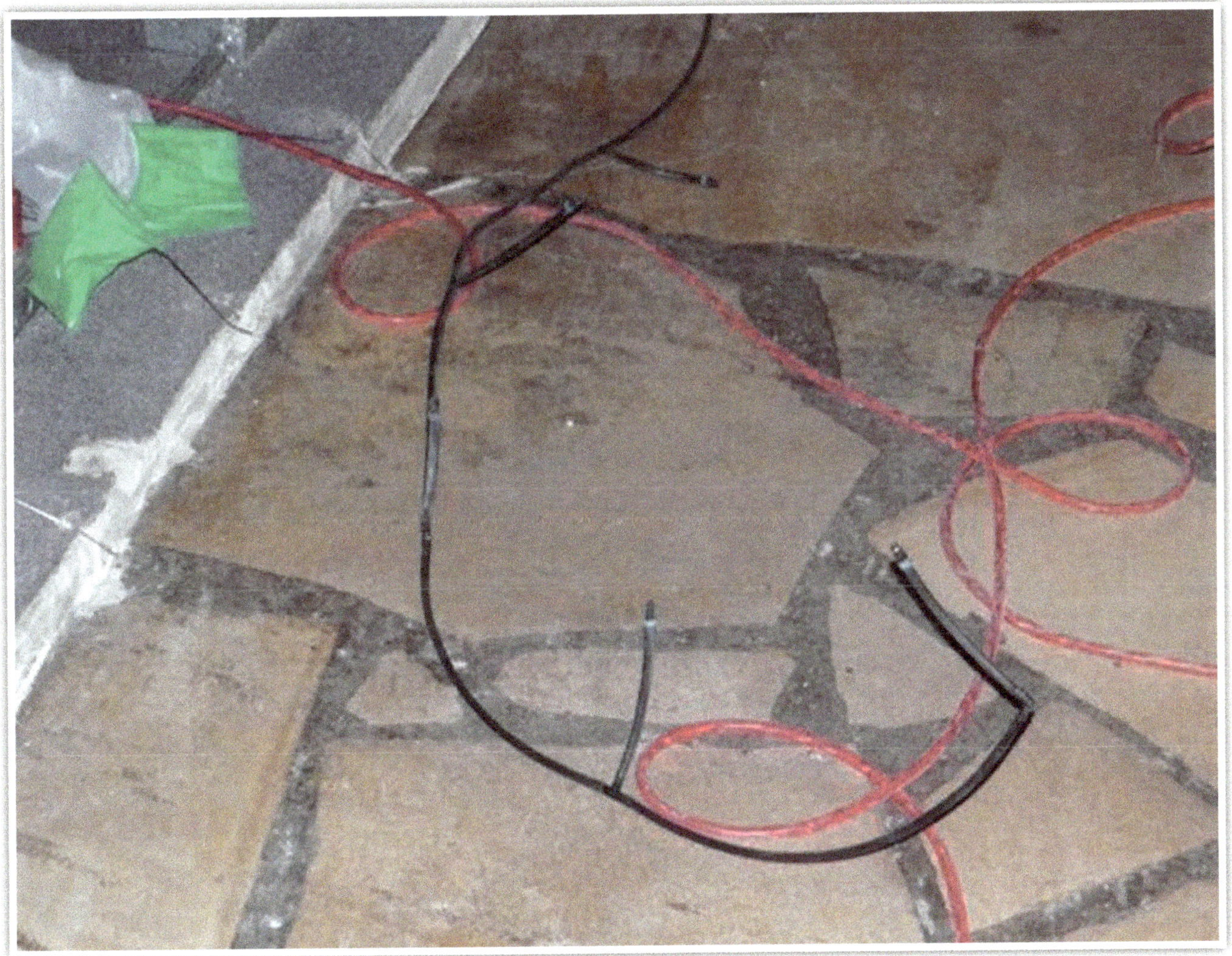

Sub-slab inject as per label

Re-install PDS ¼ inch line

Work clean again

Install PDS ¼ inch valve box and feed lead line

Secure access door

Don't forget to clean your mess and any other mess you see

There is much more to the installation of PDS systems then what is presented here

You really have to want to (see chapter 8 pictures and diagrams).

CHAPTER 9

CHALLENGES IN UNDERSTANDING THE LABEL

The dilemma you humans have is not a dilemma at all! It's just a matter of *wanting to.* Let me explain.

Words are used to convey thought patterns, and the thought patterns can result in a certain action, depending on the individual's understanding of the words. Now if a word has a number of meanings, you have a problem in trying to communicate the need for a certain action.

For example, if you were to read "walk on the side of the road," what would you do? Walk on the sidewalk or close to the curb of the sidewalk? If there is no sidewalk the answer is easy, but if there is a sidewalk you really don't know what to do unless it specifies clearly where to walk. Are you confused? Well, try this! Don't walk on the side of the road; use the public pathway. It is a violation of the law to use the side of the road. We termites have tried to understand your language so we can protect ourselves, but because of the ambiguity, *you have to want to.*

> For example, on the product label for Dragnet SFR Termiticide/Insecticide, the "Pest under Slabs" section reads as follows:
>
> Infestation of *Arthropods*, such as ants, cockroaches and scorpions inhabiting under the slab area may be controlled by drilling and injecting or horizontal rodding and then injecting 1 gallon of a 0.5% to 1% emulsion per 10 square feet or 2 gallons per 10 linear feet (italics added by Izzy.)

Please bear in mind that we termites are arthropods (see notes for chapter 9–Arthropod). Now read the next section, "Pest Control in Crawlspaces":

> Broadcast Dragnet SFR at 0.5% to all surfaces in crawlspaces to control ants, fleas, roaches, scorpions or other *arthropods*. Product may also be applied through *under-the-structure* insecticidal delivery systems such as piping or flexible tubing mounted under the structure. This treatment is *not intended as a substitute for termite control*. Treat surfaces to point of run-off. Keep children and pets off surface until dry. (italics added by Izzy.)

Now remember I said we "termites are arthropods". Do me a favor again, and you decide whether you are allowed to use the product described here in a pest control delivery system to treat termites in a crawlspace that has active evidence of my aunts and uncles. Is there any difference between this label and the instruction to walk on the side of the street? I don't think so! Thus, the real issue for an exterminator is not "am I in compliance?" but "am I doing my job to get rid of Izzy and his progeny while still being in compliance?" The answer? *You have to want to.* Also notice it says "Keep children and pets off surface until dry." This I don't understand!

NOTES

I want to remember or forget.

SECTION 3

TOMORROW (2011-????)

CHAPTER 10

SAFETY

The dictionary defines *safety* as "The condition of being safe; freedom from danger, risk or injury." With that thought in mind, the thought process of a PCO is something entirely different from that of a homeowner or a commercial building's manager. When referring to the aquifer or the soil, the rules change again; if you are referring to pedestrians, pets, or endangered species, again the rules change. This can make it complicated to execute a treatment program for Izzy and his friends that will not permit him to evacuate.

Having said that, the number one factor is that *you have to want to,* but in order to *want to* you have to know why *you have to want to.* The *why* is that life is involved. In talking about life I don't only mean a living body, because you can be in a hospital bed and on oxygen and still have life, but that is not the type of life to which we are referring. We are talking about a good quality of life, one that is full of joy and good health.

The homeowner should feel the same about the aquifer not becoming contaminated or compromised as does a PCO. Also, homeowners should want the soil to be more nutritious and to contain more organic matter and their homes to be free from contaminants and therefore passersby being in no danger, pets being in no danger, and endangered species being protected. With all the above said, though, you still have to accomplish your mission.

What do you do to be in compliance with all of the above? Did you notice that we didn't mention the regulatory agencies? That's because safety is very difficult to legislate. You can make all the laws you want, but unless you want to, you will never be safe.

It is also good to remember the long-lasting effects of good safety procedures. Just because you are not affected now does not mean you are working safely.

Sometimes, because of cross-contamination, it takes time for problems to surface. It is also good to remember that people's immune systems differ. Children's immune systems are not fully developed, and thus they experience problems that wouldn't affect adults.

Now here is the point! We could go on and on about all the different safety procedures and equipment, but this is what we say: It is not really difficult to be safe if you keep a few simple principles in mind:

"All things therefore that you want man to do to you, you also must do to them. This, in fact, is what the Law and the Prophets mean." So said Jesus in his Sermon on the Mount in Matthew 7:12 (NWTHS). With that thought in mind, how would you like to be treated? Would you want your house contaminated, your water polluted, or passersby taking home toxic waste on their clothing or their person, not to mention their lungs? (Remember, if you can smell it, you carry it!*) If you factor in the wise words of our master, Jesus, you are on your way to a safe job and environment. The results will be thus because *you did want to.*

You will now want to protect your lungs by wearing an appropriate mask and protect your eyes according to the label's recommendations. You will look forward to using appropriate rubber gloves and shoe protection. You will very quickly learn to work efficiently in your monster suit.

What do you do if you get careless and cause a spill? Well, don't you clean it and decontaminate it? Of course! Now, let me ask you: What do you do if—by error, of course!—you are contaminated? We will get the answer in the next chapter.

* And just because you can't smell it doesn't mean it's not there!

NOTES

I want to remember or forget.

CHAPTER 11

TOXINS VS. TOXEMIA

The question posed was: What do you do if—by error, of course!—you are contaminated? What are the proper procedures? Before answering that question and discussing toxins vs. toxemia we have to mention the simple truth and safety guidelines to govern any chemical procedures.

The apostle Paul in writing to the fledgling congregation in Corinth, Greece, said (as recorded in 1 Corinthians 6:12 (NWTHS), in 55 CE): "All things are lawful for me, but not all things are advantageous." In other words, just because you are right it doesn't make it right. If you approach a green traffic signal you have the right of way, but what would you do if you see in the oncoming traffic a car barreling down the highway with no apparent intention of stopping? Remember, you have the right of way legally. More than likely you would stop. Why? Because not all things are advantageous—It wouldn't be to your advantage to exercise your right.

If you recall, DDT and chlordane were considered safe for use, even in hospital rooms. At times, entire beaches were fogged with a heavy cloud while the beaches were completely occupied. Only later were health problems attributed to DDT and chlordane, as well as organophosphates.

I can't over-emphasize to you the importance of the concept that being right doesn't make it right when dealing with termites and, for that matter, in all aspects of life. Now, reaching the subject of toxins vs. toxemia you will get the answer to the question mentioned above: What if you get contaminated?

Listen in on a conversation between Jim and Izzy on one of his recent trips to see the conformed termite now trying to go straight:

JIM. Izzy ... Izzy ... where are you?

IZZY. Over here.

JIM. Where? I can't see you, but I can hear you!

IZZY. Over here in the beams. I think I consumed a piece of contaminated cellulose that contained boron.

JIM. Contaminated!

IZZY, *coughs twice*. Yes ... (*Sneezes*.)

JIM. Did you go to the termite doctor?

IZZY, *sneezes twice and then wipes his nose and antennae*. Yes.

JIM. What did he say?

IZZY. He said (*sneezes*) I have the flu. I have a temperature. I'm nauseous, and I ache all over.

Jim. What did he tell you to do? Did he give you any medicine?

Izzy. Yes. (*Sneezes twice*.) He gave me some antibiotics.

JIM. But I thought you said you consumed some contaminated cellulose?

IZZY. I did, but the doctor thinks I have the flu.

JIM. You realize, Izzy, that the symptoms for the common flu are the same as the first stage of mild pesticide poisoning and—Toxemia

IZZY. No ... I didn't know that! (*Cleans his antennae*.) What in the world is toxemia?

JIM. Rest up a bit, and we'll talk about it tomorrow. I think you'll be in a better frame of mind. What I'm going to tell you might be contrary to popular opinion, but it's well documented.

IZZY. I don't understand. (*Rubs his tummy.*)

JIM. You're not the only one! Get some rest, and I'll see you tomorrow. Stay out of that beam!

IZZY. Okay ... What in the world is he talking about? I thought the flu was just the flu. (*Sneezes twice, wipes his nose and antennae, and rubs his tummy.*) I feel terrible—I ache all over.

I hope I'm alive tomorrow. I do want to know more about that toxemia stuff. I never heard of that.

Two days later ...

JIM. Hi, Izzy. Are you feeling any better?

IZZY. Yes, much better, but I'm still sore. I'm not sneezing anymore, and I'm not shaking anymore either. What happened to you yesterday? I thought you were coming over to talk about that toxemia stuff?

IZZY. Light poisoning? ... I thought you said toxemia? (*Sneezes.*)

JIM. I did.

IZZY. Well, what is it?

JIM. Let's start from the beginning. Let's start with the definition so that we'll both be on the same page.

IZZY. Okay. (*Rubs his tummy.*).

JIM. According to the dictionary, toxins can be:

1. A poisonous substance within a living cell or organism (synthetic toxicants created by artificial processes are thus excluded).

2. Small molecules or peptides that are capable of causing disease on contact with, or absorption by, body tissues interacting with biological macromolecules, such as enzymes or cellular receptors.
3. Varied in their severity, ranging from unusually minor (such as a bee sting) to almost immediately deadly (such as botulism toxin).

by artificial processes are thus excluded as toxins?

IZZY. What?

JIM. That's right, but that doesn't mean they're not poisoned. The words *poison* and *toxin* have become synonymous with one another.

IZZY. Whoa ... You humans are sure confusing! (*Rubs his tummy.*)

JIM. And I'm not through yet.

IZZY. You mean there's more? (*Rubs his tummy.*)

JIM. Yes. (*Also rubs his tummy.*)

IZZY. Why are you rubbing your tummy?

must be sympathy pains! At any rate, let's talk about toxemia.

Toxemia is defined as blood poisoning resulting from the presence of toxins, a bacteria toxin, in the blood, says Wikipedia. It's also a generic term for the presence of toxins in the blood.

Dr. Norman Walker, in his book *Fresh Vegetables and Fruit Juices: What Is Missing in Your Body?* defines toxemia as tabolism or digestion, creating an over-acid condition." Izzy, you're falling asleep!

Does this information bore you? You know, if you understand this information you'll be better prepared to understand what goes on in your body when you're not feeling well.

Izzy, *rubbing his eyes and trying to stay awake.* I'm sorry. I know it's important, and I want to understand—go ahead.

Jim. Okay, Izzy. I'll get to the point.

When used non-technically, the term *toxins* is often applied to any toxic substance, even though the term *toxicant* would be more appropriate. Toxic substances not directly of biological origin are also termed *poisons*, and many non-technical and lifestyle journalists follow this usage to refer to toxic substances in general.

In the context of alternative medicine, the term is often used to refer to any substance claimed to cause ill health, ranging anywhere from trace amounts of pesticides to common food.

Izzy. Whoa ... I'm awake now! So that means we are surrounded by poisons?

Jim. Yes, that's right! (*Claps his hands.*) Even your negative thoughts, such as hatred, not being forgiving, or holding a grudge, can cause poisons in your body. Can you understand why we have so many health clinics and hospitals?

Izzy, *scratching his head.* I'm sure glad I stayed awake most of the time.

Izzy: W... O... W...! All that's Toxemia?

Oh no, replies Jim, what I just said has to do with contamination, self-contamination, nothing to do with chemicals!

You see Izzy, if you don't take care of your vital organs by keeping them clean, it's the same as working without your protective equipment. So, a regular self-cleaning program is in order, or to put it in plain English, you should detox on a regular schedule. Richard Anderson, N.D., N. M.D., recommends at least four times a year, once every season, in his book CLEANSE & PURIFY THYSELF. If you work with chemicals at least every month. Paul C. Braggs and his daughter Patricia recommends in his book THE MIRACLE OF FASTING, too fast for twenty-four hours once a week and for thirty-six hours once a month. That's three twenty fours and one thirty-six a month. Believe me Izzy it makes you feel better, and you rid yourself of many toxins: both chemicals and produced, and your organs will love you for it.

Izzy: Boy Jim, you are changing my whole way of viewing safety.

Jim: What do you mean?

Izzy: Well, according to what you said, if we don't clean our vital organs you could develop the same symptoms as if you worked with chemicals.

Jim: That's right!

Izzy: so, what's the best way of protecting yourself?

Jim: O. K. Izzy let me give you an illustration.

Izzy: O. K.

Jim: What do you do if your plumbing in your home is plugged, or better yet - what happens when your main drain is plugged?

Izzy: Well....If your main drain is plugged that means all your drains are plugged.

Jim: That's right! and what would happen if you continued to use it?

Izzy: I guess it will start to smell and even overflow.

Jim: Thats's right, could you imagine if you continued to use your toilet?

Izzy: EEEEEEEKKK

Jim: Now Izzy! Pay close attention to what I say next.

Izzy: I'm listening.

JIM: If you unplug the main, that is the main sewer lines all the drains will flush and empty; thus, removing all poison from your drains, we say poison because when food particles and digested waste are stagnant in pipes full of water, they grow deadly bacteria and that my friend is—Poison!

IZZY: Cool....I'm sorry I don't mean good; I mean that's interesting and valuable information, but my plumbing is not plugged.

JIM: No, I know, but if your colon is plugged or partially closed your whole organ system is being partially plugged, or closed, putting stress on all your organs, and so they begin to atrophy because of not getting the nutrition they vitally need. So, like your plumbing you have to unplug your colon. I know it's not the most popular subject to talk about but you must address the health of your colon if you wish to be healthy.

IZZY: I think I'm starting to get it. See if I'm right! O. K. says Jim, when I empty my colon toxins in my organs now have a place to empty into because I pulled the plug and now they have room to empty and continue to function to capacity.

JIM: Hurrah, Hurrah Izzy now you can say, cool!

IZZY: but Jim, you never talked about water as you said you would!

JIM: Oh, you're right. Thanks for the reminder. Before I talk about water just one more thing.

IZZY: Here we go again!

JIM: No....No....this will be short. I'm just going to give you the signs of an unclean colon.

1. Bad breath
2. Bad body odor
3. Bad or Putrid Smell of fecal matter
4. Flatulence-Gas
5. Acne
6. Excess mucus in head, throat, lungs
7. Kidney Infection
8. Prostate Problems etc.
9. Female Problems
10. Liver—Gall Bladder
11. Chronic Illness

Izzy: How do I clean my colon?

Jim: I'm not a doctor, that is of human health so all I can say is to read up on the subject and keep an open mind and don't go to extremes. Enjoy your life, we have very good things to look forward to... Rev. 21:3,4.

I gave you some references at the back of the book bibliography.

Jim, *walking across the room toward the bookshelf.* Are you up to hearing some more?

Izzy, *smiling*. You mean there is still more?

Jim. Well, yes, but I'll be brief. I'm going to talk about the heart, liver, gall bladder, pancreas, colon, and water in the body.

1.

(*Stands in front of the bookcase, pulls out the book* School of Arthropod Barrier Construction Engineering, *and opens it to the chapter on health and safety.*)

Izzy, *a big, happy grin on his face*. I'm sure glad I'm an arthropod and not a human.

Jim. (*Laughs.*) Let me show you this.

The information given in this presentation is not medical advice and is not given as medical advice, nor is it intended to propose, or offer to propose, a cure for any disease or condition.

Before beginning any medical treatment, please consult your physician.

The Heart

- How does it work
- Why does it work

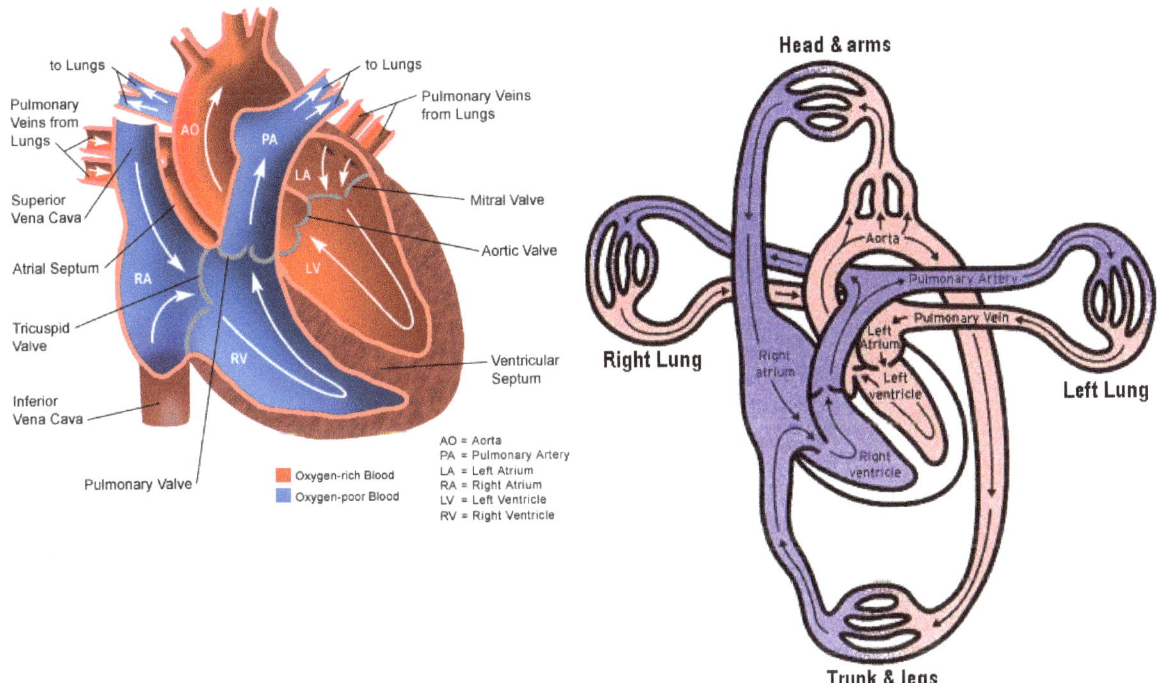

The heart is purely and simply an automatic organ. The automatic beat of the heart results from the function of both the muscles and the nerves with which the heart is equipped. however, these alone do not contain the activating force which keeps the normal, healthy heart striking some 100,000 beats each day for 50, 75 100 years and more; yet this miraculous pump is just about the size of a man's fist. The blood is the life of the body, while the heart is the motor that keeps that life in circulation. The entire body contains only about five quarts of blood. This is its entire supply of blood and a healthy body does not add any more to it throughout its entire life. nevertheless, during each 24-hour period, day after day, year each 24-hour period, day after day, year in and year out, this tiny pump, the heart.

Some 45 million gallons of blood through your system. Miraculous? No human could contrive a device that could begin to equal that performance! The heart is very definitely affected by impurities in the blood—by fermentation and putrefaction of waste matter anywhere in the system—because these impurities are collected by both the blood and the lymph. The blood and the lymph fluid passes through the heart constantly. Gases from the colon can readily pass into any part of the body by the process os gas-osmosis. Obviously, when a pocket of gas forms anywhere in the heart area, it is very likely to cause trouble, whether the gas is inside the transverse colon or in the area of the diaphragm.

The Liver

In our spray rings we call it the filter.

And It's companion

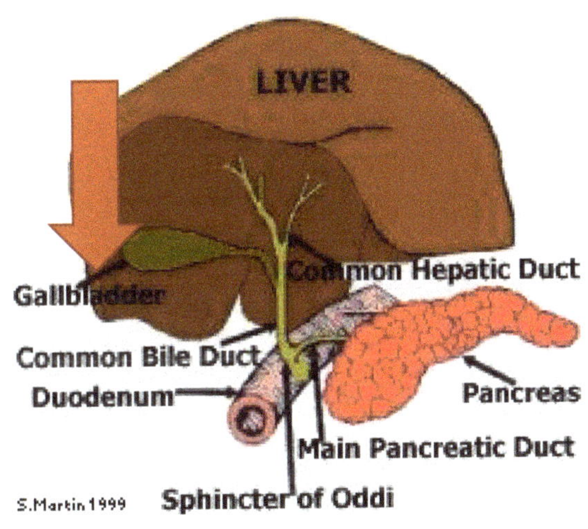

At the top of the ascending colon, where it curves to the right on the chart (curves to left in the body) is the area marked liver. This curve is known as the hepatic (or liver) flexure, and is located immediately above the liver. The liver is composed of the right lobe and the left lobe, the right lobe being by far the greater and larger of the two. The liver is the largest gland in the body and is the organ carrying on the most active and extensive operation in the anatomy. Everything that goes down the throat, (in the form of food or otherwise, and every beverage that we drink, sip or pour down our throat), must pass through, (in more or less liquid form), the 25 feet of small intestine. The small intestine is equipped with millions of tiny suc-material inside the small intestine and pass these **molecules** on to the surrounding blood vessels. The blood in these vessels instantly takes these **molecules** to the liver where they are split up into their component atoms. the segregated atoms are then catalogued and assigned to other atoms which, in turn, form new **molecules** of the kind the body calls for and which the cells and tissues of the body can use.

Thus, it is useless to eat such things as "a complete protein" with the expectation and the false assurance that this complete protein will be used by the body. The so-called complete protein must first be emulsified into a heterogenous mass called *chyme* and mixed with everything else in the small intestine. In such a state all molecules in the *chyme*, whatever they happen to be, are gathered by the

What was originally sugar and starch is likewise broken down into their respective molecules and these, in turn, are disintegrated into the separate atoms composing them and reassembled to form glucose. It is then converted into *clycogen* and stored for reconversion into glucose at which point it can be released into the bloodstream for delivery where it is needed. Glucose is a form of sugar present in many fruits; for instance grapes, apples, bananas, and also in honey

It can be found in the lymphatic system and the bloodstream. Glycogen is a tasteless carbohydrate related to dextrin and starch.

A very important activity of the liver is the generation of bile. After bile is generated it is stored in the gallbladder to be released into the duodenum, (or second stomach), whenever anything passes through it from the stomach. Suffice it here to say that bile is involved in the digestion and absorption of fats.

The liver is greatly concerned with preventing blood-clotting. Other functions of the liver are concerned with fat and protein metabolism. The liver is a detoxifying agent and a blood reservoir. It breaks down the hemoglobin of the red blood cells, which have already served their purpose, and it also stores copper, iron and other trace elements for instant use when needed. If you nourish your body properly, keep your colon clean and your mind elevated, your liver will take.

The Gall Bladder

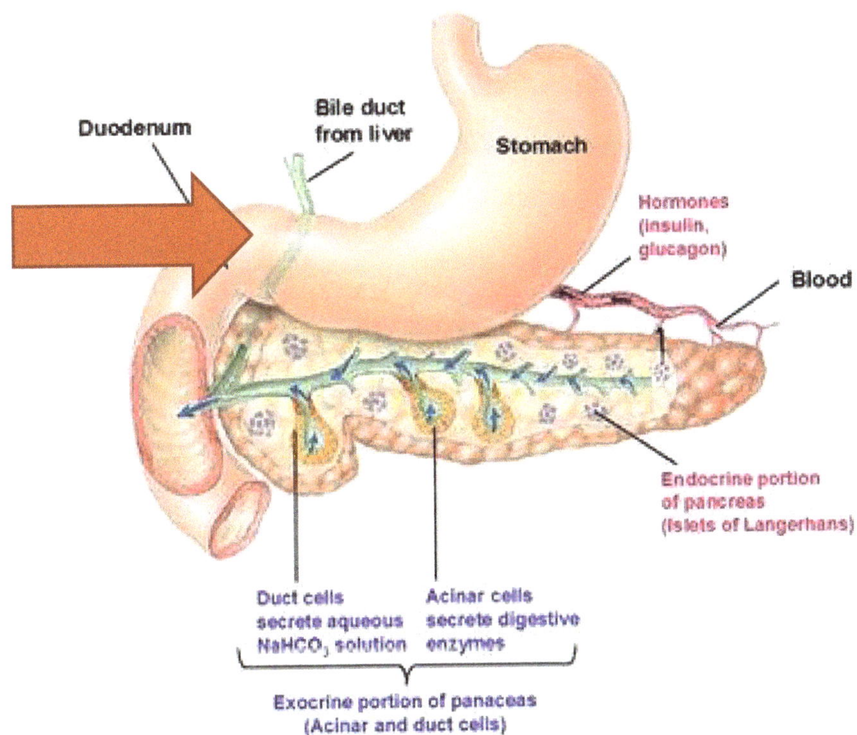

The gall bladder is a bulbous sack attached to the underside of the liver in order to serve as storage for the bile secreted by the liver.

Bile is an important element in the process of digestion as it is mainly involved in the digestion of fats. Bile helps in the digestive processes by neutralizing the acid *chyme*, which passes through the duodenum from the stomach by emulsifying fats. It also promotes peristalsis, the absorption of elements in the body, by helping to prevent putrefaction.

Bile is composed of certain acids which form an intimately-related group of natural products. While bile acids occur free to some ex- Or taurine (as in glycocholic acid and tau rocholic acid), which results from the breakdown of proteins. Two of the better known bile acids are colic and lithocholic.

Cholesterin (cholesterol) is a white, fatty, crystalline alcohol, tasteless and without odor, found abundantly in the tissue of nerves. It is also present in bile and in gallstones. It does not contain nitrogen so it is not a part of protein. Lecithin and cholesterin are usually found together, which indicates a physiological aspect to their function.

The gall bladder is an extremely important organ in the body. Its removal only results in the development of subsequent difficulties in digesting food properly.

The Pancreas

- Healthy
- On Healthy

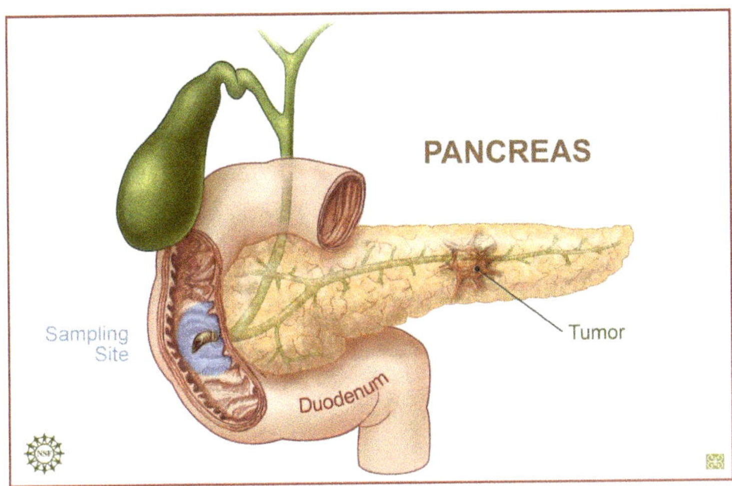

The pancreas is a long, narrow gland of internal and external secretion. It stretches from the spleen to slightly higher than the middle of the sem-circular bend of the duodenum. Its main duct joins the duct of the gallbladder. Emptying into the duodenum, the pancreas is a compount tubular gland like the salivary glands in the mouth.

The pancreas is one of the most active glands in the system. Everything that goes through the duodenum requires some of the digestive juices produced in the pancreas. Pancreatic juice contains digestive enzymes and is alkaline in its reaction, serving to estab.

Toward the middle of the pancreas there is a group of glands of internal secretion. They are known as the islands of langerhans, which produce insulin, the hormone that regulates the metabolism of sugar (the blood sugar level) and other carbohydrates. When the body is toxic and the colon is afflicted with fermentation and putrefaction, these glands are unable to produce this insulin, causing an intolerance to sugar in the body. Under these circumstances the volume of sugar is increased in the blood and is discharged into the kidneys. This condition is called sugar diabetes or diabetes mellitus. Many people so afflicted have found that the cleansing of the colon by means of a series of colon irrigations, coupled with a complete change of diet of fresh, raw vegetables, fruits and fruit juices. Additionally, nuts and sprouted seeds

They found they derived much benefit from drinking one or two pints daily of the combination of carrot, celery, string bean and brussels sprout juices.

The pancreas is a very important gland in our body and we should give to it the full complement of respect which it deserves if we wish to attain any degree of perfect, vibrant.

Why am I Sick ?

" I shall laud you because in a fear-inspiring way I am wonderfully made. your works are wonderfu, as my soul is very well aware " PS 139:14

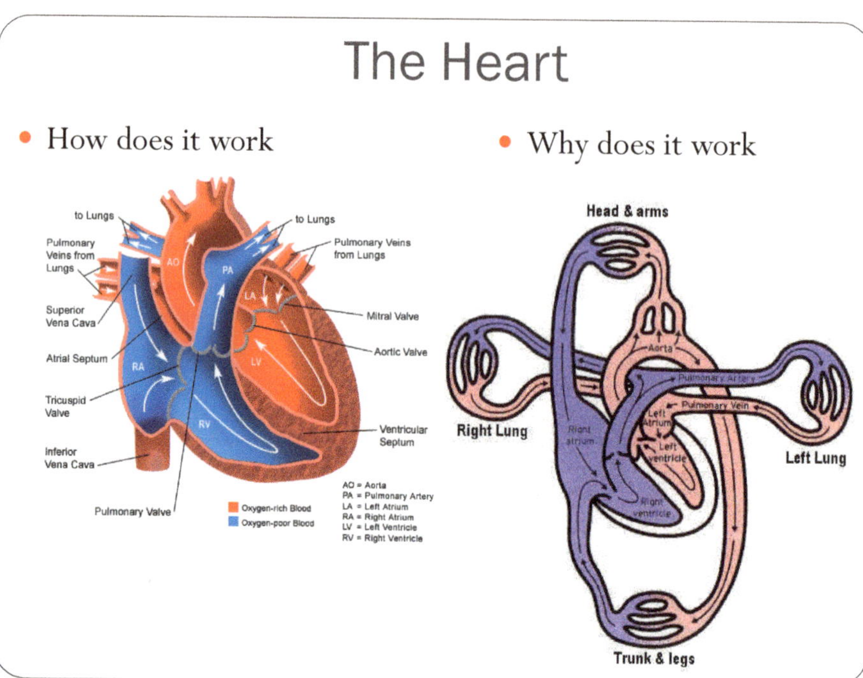

Where does the blood go and why?

- It goes to the lungs first to be oxygenated

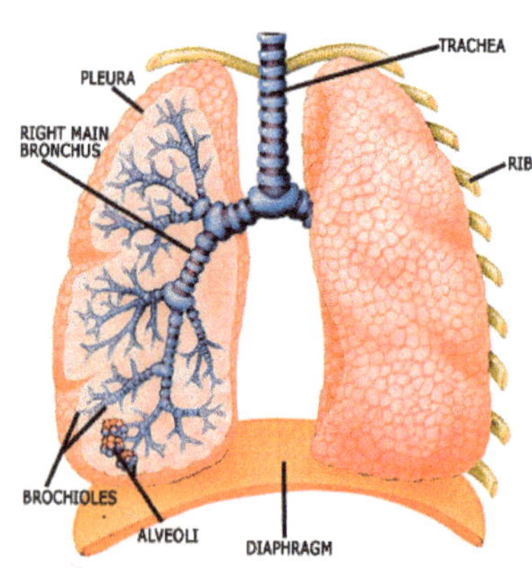

Imagine if you would, what happens if your blood is carrying toxins of any kind be it chemicals or toxins produced - do to a unhealthy diet!

Not The End

The

Beginnings

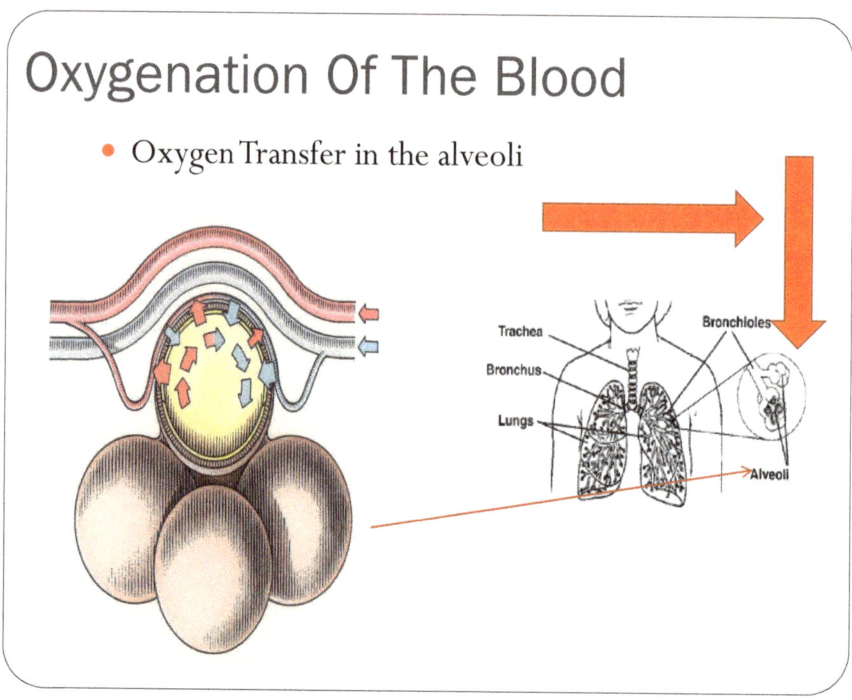
Oxygenation Of The Blood
• Oxygen Transfer in the alveoli
Trachea
Bronchus
Lungs
Bronchioles
Alveoli

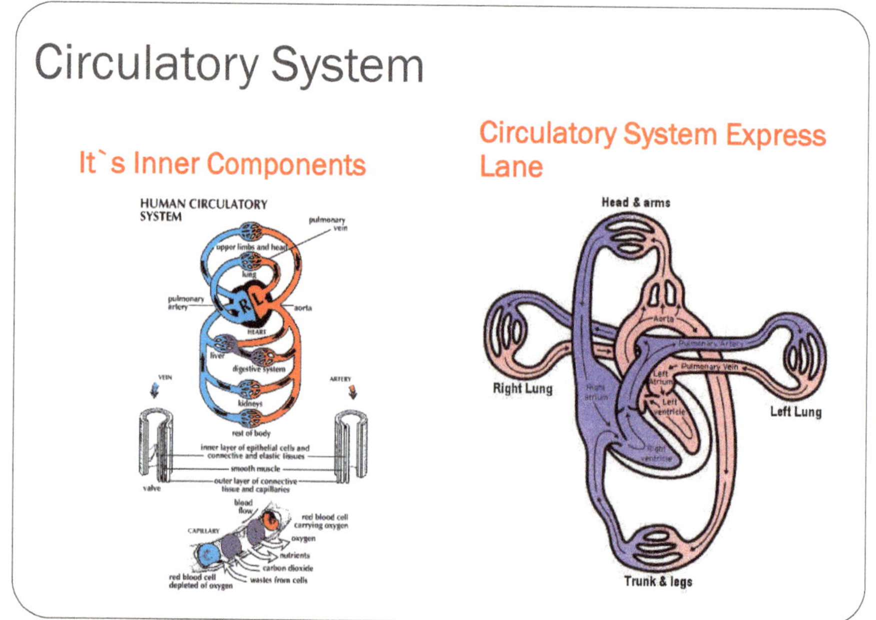
Circulatory System
It`s Inner Components
HUMAN CIRCULATORY SYSTEM
pulmonary vein
upper limbs and head
lung
pulmonary artery
aorta
HEART
liver
digestive system
VEIN
ARTERY
kidneys
rest of body
inner layer of epithelial cells and connective and elastic tissues
smooth muscle
outer layer of connective tissue and capillaries
valve
blood flow
red blood cell carrying oxygen
CAPILLARY
oxygen
nutrients
carbon dioxide
wastes from cells
red blood cell depleted of oxygen
Circulatory System Express Lane
Head & arms
Aorta
Pulmonary Artery
Pulmonary Vein
Right Lung
Left Lung
Trunk & legs

Route of oxygenated Blood

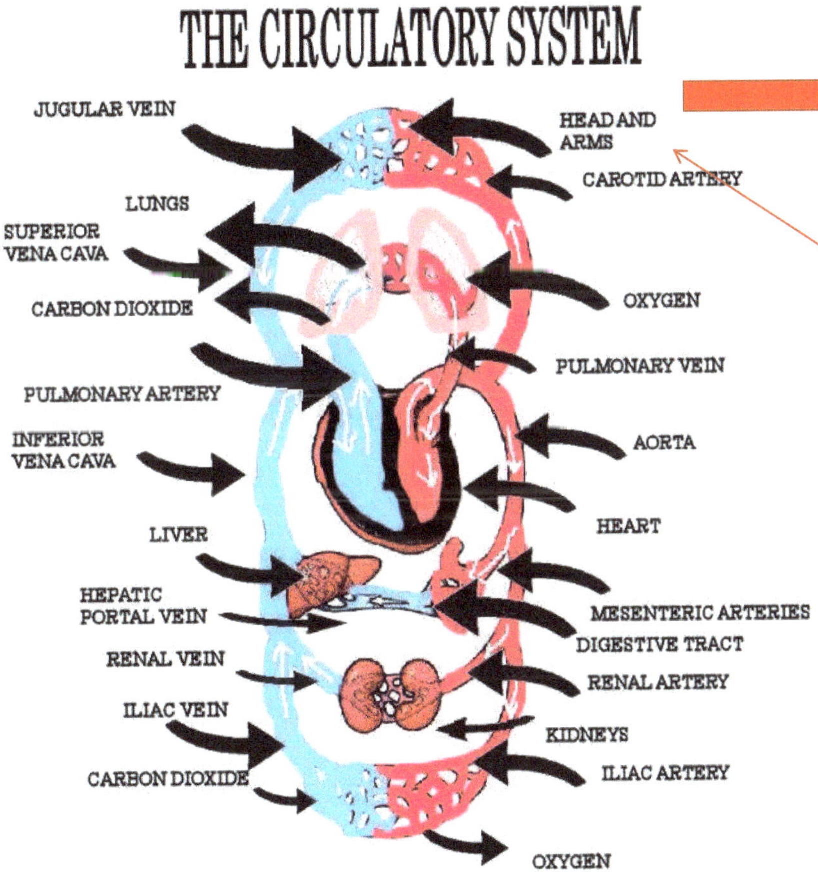

THE CIRCULATORY SYSTEM

JUGULAR VEIN
HEAD AND ARMS
CAROTID ARTERY
LUNGS
SUPERIOR VENA CAVA
CARBON DIOXIDE
OXYGEN
PULMONARY VEIN
PULMONARY ARTERY
INFERIOR VENA CAVA
AORTA
HEART
LIVER
HEPATIC PORTAL VEIN
MESENTERIC ARTERIES
DIGESTIVE TRACT
RENAL VEIN
RENAL ARTERY
ILIAC VEIN
KIDNEYS
CARBON DIOXIDE
ILIAC ARTERY
OXYGEN

Or contaminated blood

Have you ever said, " I have a headache" ?

The Liver

1-The Liver is the largest organ in the body
2-Over 5000 chemical Functions
3-It can clone it self, if part is removed
4-It's byproduct is bile, used to help in the digestion of fat

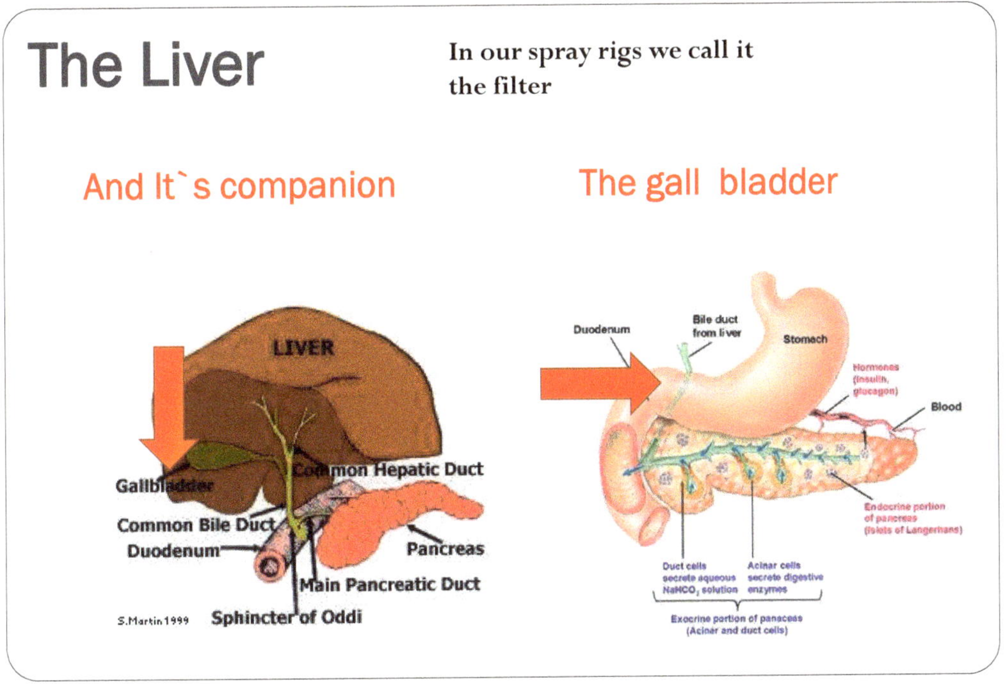
The Liver
In our spray rigs we call it
the filter
And It`s companion
The gall bladder
LIVER
Gallbladder
Common Hepatic Duct
Common Bile Duct
Duodenum
Pancreas
Main Pancreatic Duct
Sphincter of Oddi
S.Martin 1999
Duodenum
Bile duct
from liver
Stomach
Hormones
(insulin,
glucagon)
Blood
Endocrine portion
of pancreas
(islets of Langerhans)
Duct cells
secrete aqueous
NaHCO3 solution
Acinar cells
secrete digestive
enzymes
Exocrine portion of panaceas
(Acinar and duct cells)

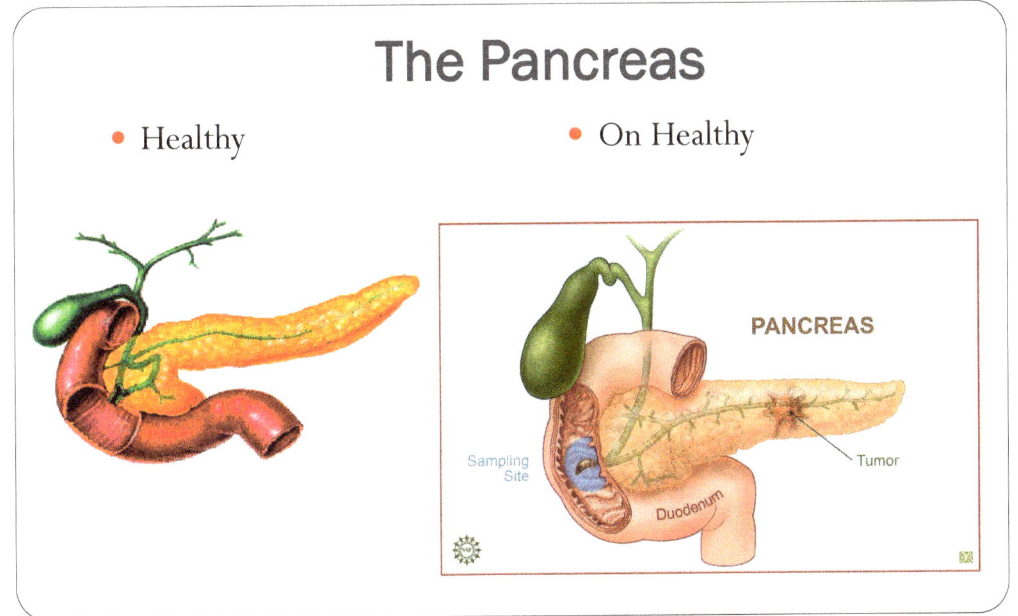
The Pancreas
Healthy
On Healthy
PANCREAS
Tumor
Sampling
Site
Duodenum

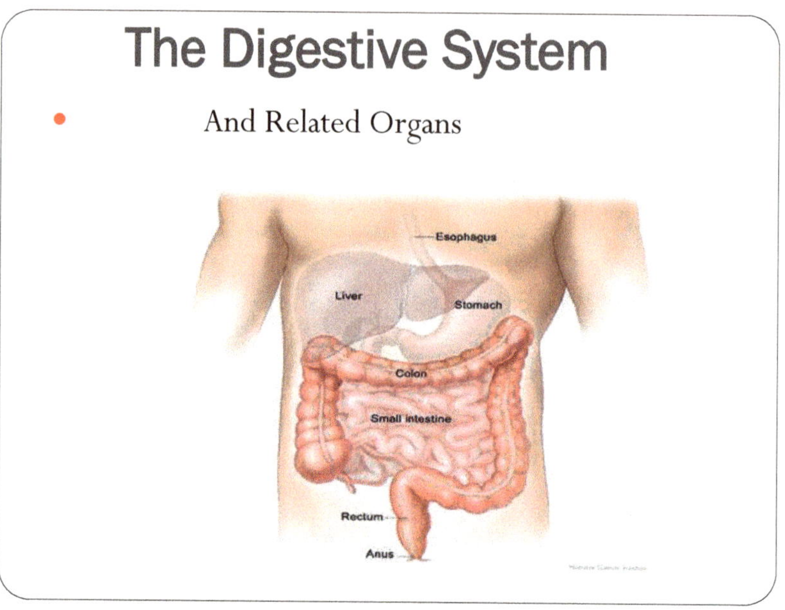
The Digestive System
• And Related Organs
Esophagus
Liver
Stomach
Colon
Small intestine
Rectum
Anus

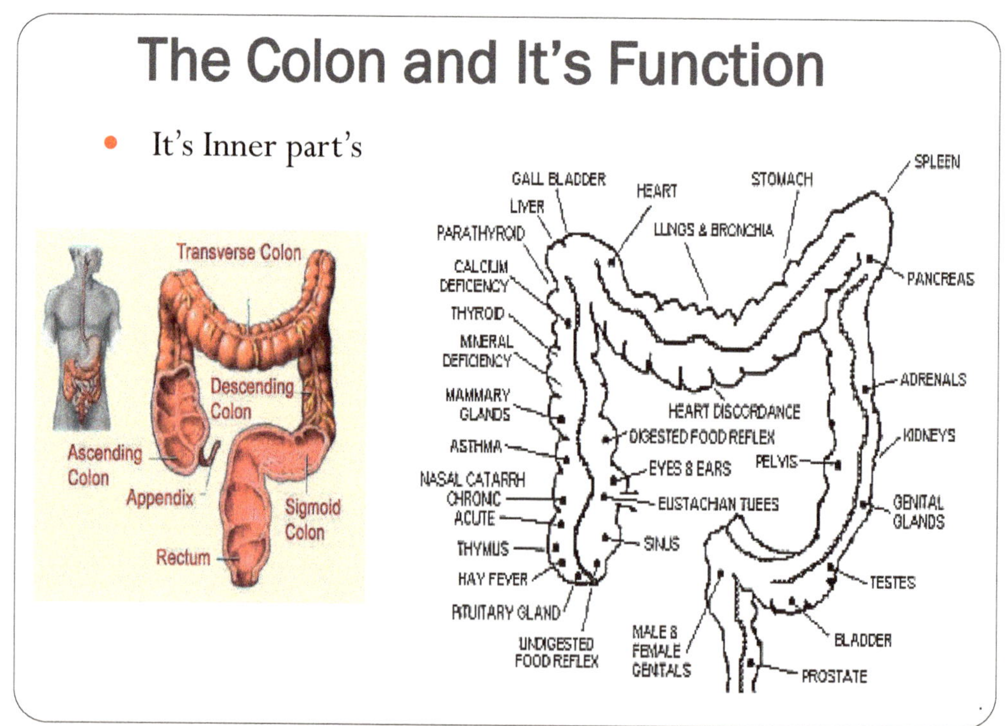
The Colon and It's Function
• It's Inner part's
Transverse Colon
Descending Colon
Ascending Colon
Appendix
Sigmoid Colon
Rectum
GALL BLADDER
LIVER
PARATHYROID
CALCIUM DEFICENCY
THYROID
MINERAL DEFICIENCY
MAMMARY GLANDS
ASTHMA
NASAL CATARRH
CHRONIC
ACUTE
THYMUS
HAY FEVER
PITUITARY GLAND
UNDIGESTED FOOD REFLEX
HEART
LUNGS & BRONCHIA
STOMACH
SPLEEN
PANCREAS
ADRENALS
KIDNEYS
GENITAL GLANDS
TESTES
BLADDER
PROSTATE
MALE & FEMALE GENITALS
HEART DISCORDANCE
DIGESTED FOOD REFLEX
EYES & EARS
PELVIS
EUSTACHIAN TUBES
SINUS

The Colon and Tree roots Are Related In Function

- Different parts of the body are feed by designated parts of the colon
- The upper part of the tree is fed by the lower part-the roots

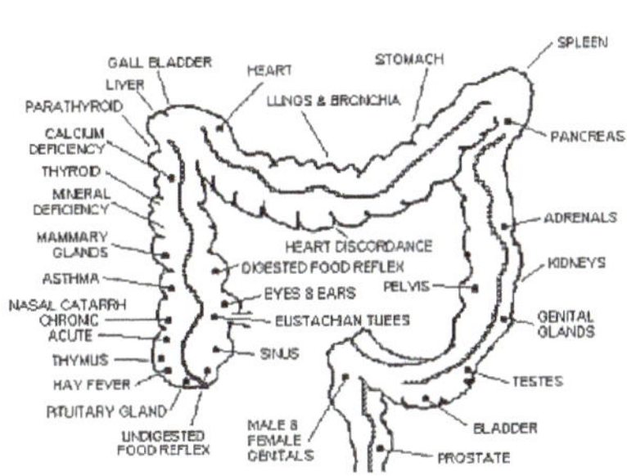

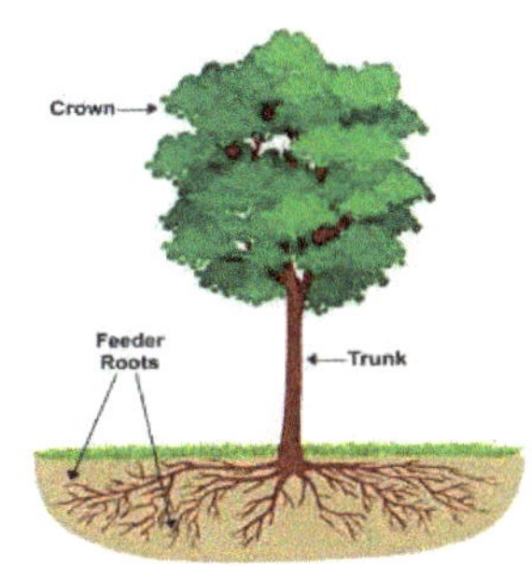

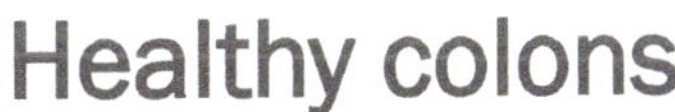

Healthy colons

Sickness

Healthy People

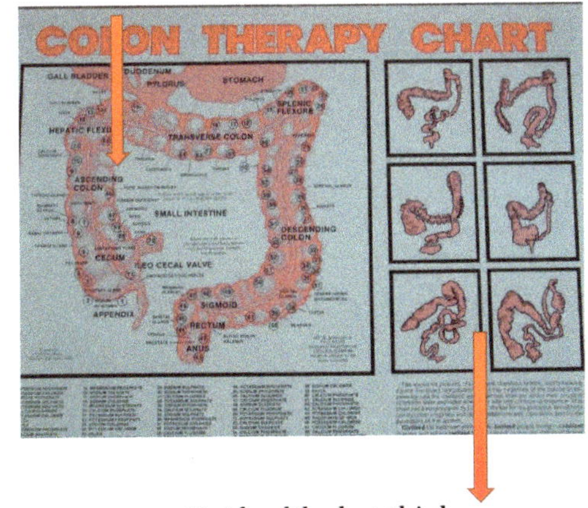

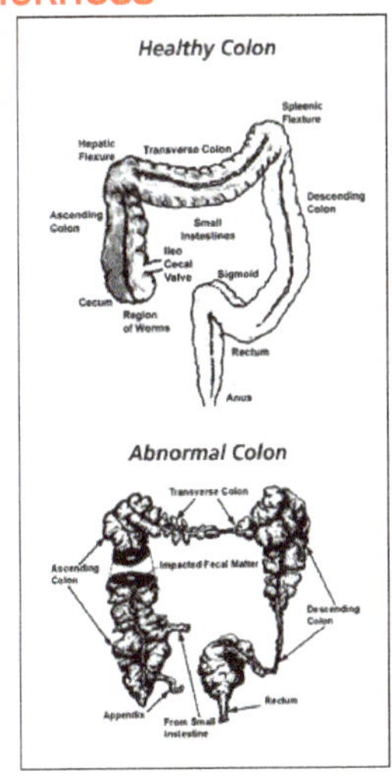

Not healthy but think they are

What happens if the colon is plugged

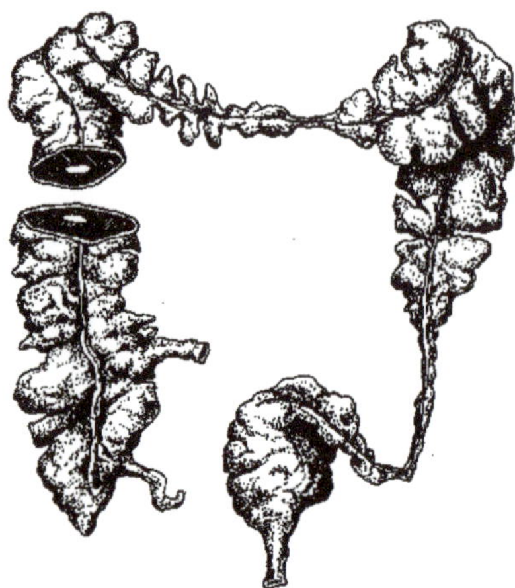

It's like cutting the roots of a tree, the affects are seen in the crown

Disease is generally not acquired; it is frequently created

If we allow our bodies to become Acidic and congested

Clogged spray pump

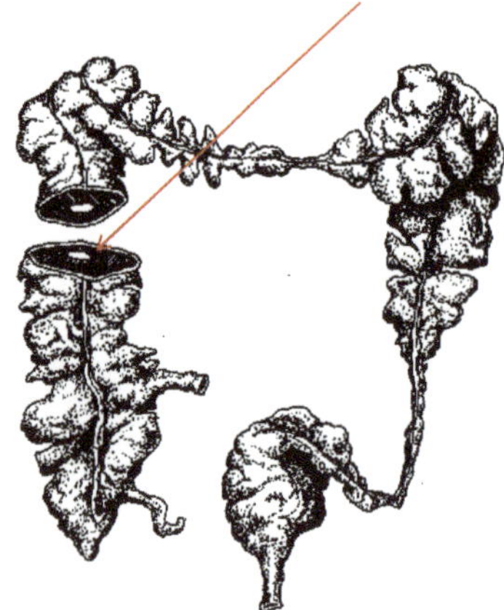

WE RUN THE RISK OF WEAKINING OUR GLANDS AND ORGANS

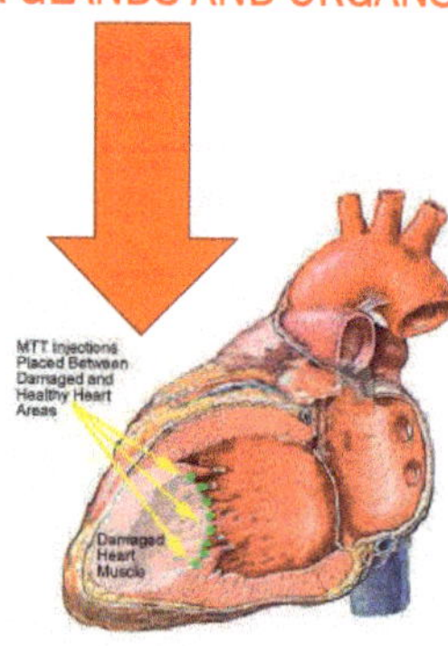

It's like running acid through our spray pumps

A WEAKENED PANCREAS, OR UNDER STRESS

Do to improper nourishment or holding on to grudges and hateful feelings for a prolonged time, the same for all our organs.

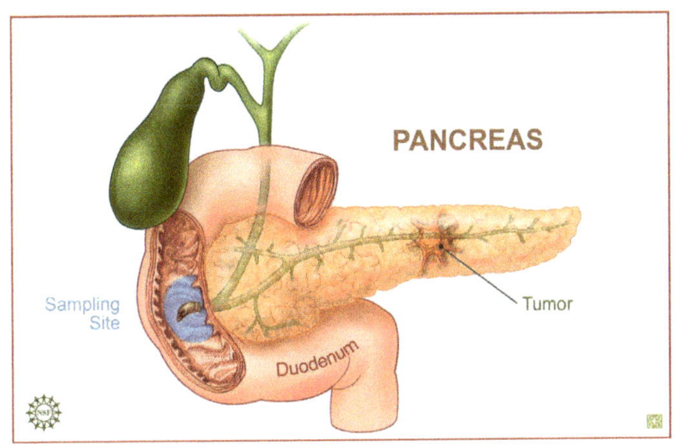

Lung with emphysema

If your hose is plugged, nothing works

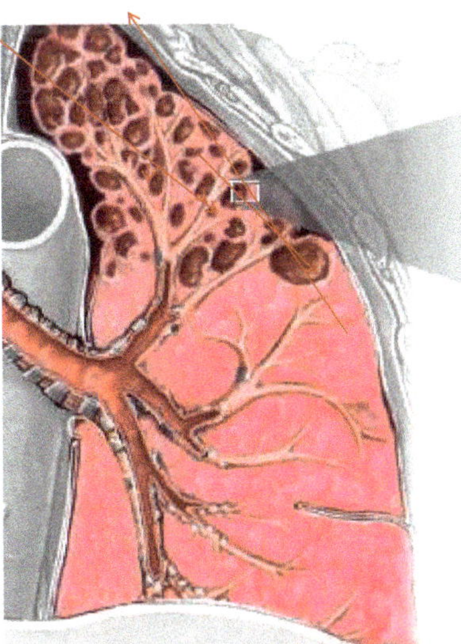

Alveoli with emphysema

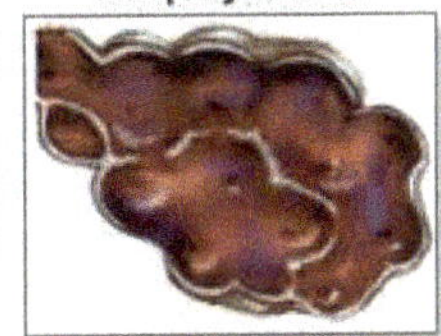

Microscopic view of normal alveoli

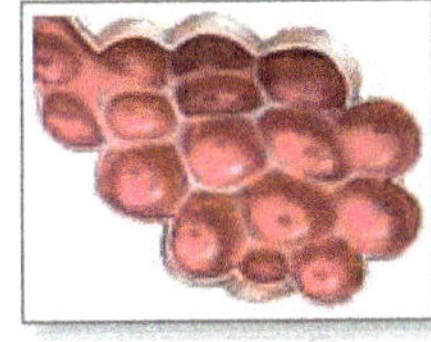

What is the function of axons and dendrites?

- Dendrites receive (take in) signals from other neurons, or from sensory cells (cells that give information about sight, sound, smell, taste, touch).
- Axons transmit (send) signals to other neurons or to other cells in the body.

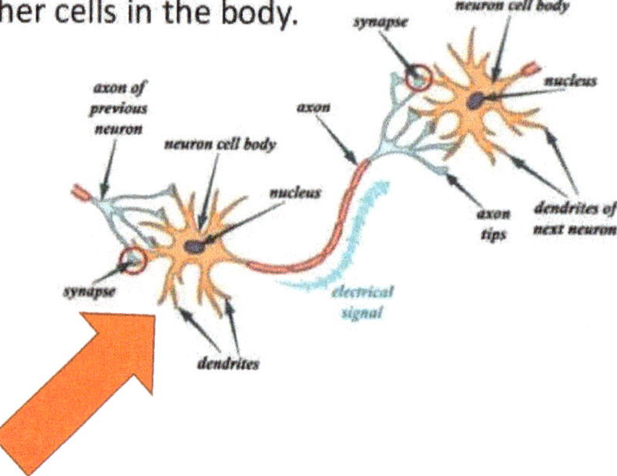

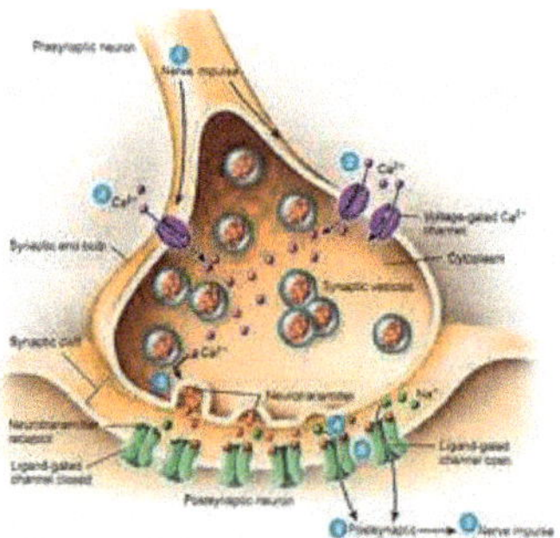

A synapse is a **small gap at the end of a neuron that allows a signal to pass from one neuron to the next**. Neurons are cells that transmit information between your brain and other parts of the central nervous system. Synapses are found where neurons connect with other neurons.

CHAPTER 12

TOMORROW

When we begin talking about tomorrow and what it will cause for us, one must look to the development of the past and its progression toward change. With that pertinent information we will be able to look to the future with almost prophetic insight. For example, the forties, fifties, and sixties were the heyday of CHC use.

However, with the development of science and health, the thinking of the public began changing, thus putting CHCs in a not-so-favorable light. This came at a time when the termite control market was growing because of the housing boom and the need to prevent Izzy and his family from taking control of our homes. As science and health changed, so our industry changed, and by the mid-eighties CHCs were a thing of the past.

With those thoughts in mind, look at the influence of the media on our society. One might say that the media rules! Why do we say that? Well, what role did the media play in the retirement of the CHC? With the development of computers and their rapid changes came the rapid changes in the thinking of society, and the media made it known that CHCs were dangerous to society. The danger, however, was not because of their chemical makeup* but because of the slovenly application of CHCs and their over-application.

We were applying ten gallons per ten linear feet of label-rate chemical when we could have prevailed with half the amount, as the eighties and nineties proved. We then abandoned a good product but maintained the old techniques.

In talking about the old techniques, we tried using old CHC-label phraseology with the organophosphates of the eighties and nineties, but that didn't work because those chemicals didn't have the longevity of CHCs.

Again, muddled application and misapplication of the label rates (watered-down chemicals) occurred, causing new problems; CHCs had been very forgiving when it came to label rate. Bear in mind that we are trying to control termites, not sophisticated mammals like bears, cats, or rhinos.

That led to a change in marketing approach by chemical and pest-management companies. Chemical companies now target the DIY and YouTube populations. The result? Monetary gain for the chemical companies and www.problems for the DIY and YouTube audience (www.problem = What Went Wrong problem).

To illustrate: many common medical problems are compounded by DIY remedies, which flood the market (see chapter 10). These remedies, although they have some value, do not get to the real cause, and people often end up needing to call a pro, or a doctor—"the case now being that because of the lapsed time, the procedures are more complex and expensive, but they must be executed for the benefit of your health and that of your loved ones."

Now compare that with a house or commercial building that shows evidence of minor termite infestation.

What does the landlord do? YouTube and DIY to the rescue! However, if he or she is not familiar with the biology and species of Izzy and his family, the landlord might complicate the problem, making it worse and subsequently requiring not a YouTube graduate but an industry graduate, one who has the proper education and experience and the appropriate letters after his name indicates, as in James Raymond, DTBT—Doctor of Termite Biology and Treatment. This designation would be awarded by state agencies, schools, or colleges after the student has fulfilled the requirements and experience as set forth; the results will be a higher caliber of standard and procedures. With that thought in mind, what are some of the procedures we could look forward to?

We could still use tools we are familiar with: rodding, subslab injection, trenching, wood treatment, physical barriers (as in wire mesh), bait stations and monitoring, and ence of the environmentalists, research connecting the use of chemicals and the increase of diseases like ADD, plus the media coverage, we find ourselves repeating what contributed to the retiring of CHCs.

The difference is that with a more intelligent approach by our industry—and seeing the "handwriting on the wall"—we are becoming more prudent in our chemical formulation and even more prudent in our application techniques. Take for instance the label of Dupont's Altriset Termiticide. It has no warning or caution label, no specific first aid requirements, and no specification of protective equipment needed when applying the product.

Prudent indeed! It also states, in section #22.3 of the label, the following:

> Fixed SUB SLAB DELIVERY SYSTEM
>
> Permanently installed piping or flexible tubing, may also be used to deliver product to critical inaccessible areas under the slab, such as concrete expansion joints, cracks, plumbing utility services, penetrating the slab, etc. Follow manufacturer's directions for use of the delivery system to ensure that the insecticide is distributed evenly throughout the treatment zone. For these systems, the finished solution of "Altriset" Termiticide must be applied at the rate of 1 gallon per 10 square feet.

Did you notice the instructions indicate following the manufacturer's direction for the use of a delivery system? That must mean that in the near future there will be training on delivery systems, because if one does not understand the hydrodynamics of a delivery system it will be like wiring a house without understanding volts, amps, and watts. The results? www.problems.

We can predict the future from analyzing the past, when we were educated in the calibration of our tanks, nozzles, and nozzle size, as well as the speed at which we travel when making applications to acreage and the pressure for a termite pre-treat application. Recall we said: "Today is yesterday's tomorrow." Well, the above Altriset label is a good example of future treatments for controlling Izzy and his family. New labels are well underway with many changes in view. Take a look at a two-day program for the School of Arthropod Barrier Construction Engineering:

School of Arthropod Barrier Construction Engineering

Two-Day Agenda

First Day

8:00–9:00	Registration, Continental Breakfast, Equipment Display
9:00–9:15	Welcome, Introduction and short history of Pest Control Delivery System and College of Eastern Utah, San Juan Campus, by Calvin Hunt, Director of Applied Technology Education
9:15–9:45	Safety
	-Practical Pest Control Safety
	-Protect the Applicator
9:45–10:15	Identification of Wood-destroying Organisms
10:15–10:30	Break (refreshments provided)
10:30–11:30	Health and Safety
	-Toxins vs. Toxemia – A close look at the liver, gall bladder, and pancreas
	-Crawl space
	-Floating slab
	-Foundation construction
10:45–11:00	Break (refreshments provided)
11:00–11:30	Hands-on presentation
11:30–12:00	Paperwork, licensing, and pictures
12:00	Conclusion

Note: New Products Developers, Inc., personnel will remain as long as needed for questions, discussion, etc.

Jim. Izzy, why don't you take over?

Izzy. Okay. How do I start?

Jim. Just tell them what you told me.

Izzy. Okay. (*Approaches the lectern.*) Are you sure? I'm just a termite, you know. Will they listen to me?

Jim, *walking across the floor to the lectern and pointing to his notes*. The information will speak for you.

Izzy. Okay, here goes.

An important principle in all walks of life and in *whatever you do is to remember "you have to want to,"* which means you'll do it so that when the task is completed, all will know that you really wanted to, and you'll do it with *a positive and affirmative attitude.*

Search wherever you will for a single sound argument against this principle, and you will not find it; nor will you find a single instance of enduring success that was not attained in part by its application. Go the extra mile.

Consider Jesus's words in Matthew 5:41 (NWTHS): "and if someone in authority compels you into service for one mile, go with him two miles." So you can see it's not a new concept. Let us examine some of them and be convinced. Using this principle to guide your actions does the following:

- It leads to mental growth and to physical skill and perfection in many forms of endeavor.
- It develops the important quality of personal initiative.
- It develops courage.
- It serves to build the confidence of others in one's integrity.
- It aids the mastery of the destructive habit of procrastination.

Let us now observe that the admonition to render more service and better service than that for which one is paid is paradoxical because *it is impossible for anyone to render such service without receiving appropriate compensation.* The compensation may come in many forms and from many different sources, but come it will. The Bible puts it this way: "Cast your bread on the waters, for after many days you will find it again" (Ecclesiastes 11:1, NWTHS). The worker who renders this type of service may not always receive appropriate compensation from the person to whom he renders the service, but this habit will attract to him many opportunities for self-advancement. Thus his pay will come to him indirectly. Ecclesiastes 11:6 (NWTHS) continues, saying "Sow your seed in the morning and do not let your hand rest until the evening; for you do not know which will have success, whether this one or that one, or whether they both will."

"If you serve an ungrateful master, serve him more. Put God in your dept. Every stroke shall be, the better for you; for compound interest on compound interest is the rate and usage of this exchequer."

This is an asset of which no worker can be cheated, no matter how selfish or greedy his immediate employer may be. It is the "compound interest" that Emerson mentioned.

Izzy. What do you think, Jim? Did I do okay?

Jim. Yes, you did, but now you have to give credit to the Napoleon Hill Foundation for the information you just quoted.

Izzy. Oh, yes, that's right.

The information just quoted was taken in part from an essay entitled "Go the Extra Mile" by Napoleon Hill.

You may obtain a full copy of this work and other works of his by contacting the address below or calling the number provided.

1 College Avenue
Wise, VA 24293-4400
Phone: 276.328.6700

Oh Did I Tell You

That we are so special that they give us the prestigious title- **ENGINEERS-**`and, **INGENIOUS** at that. We love it, let me ex- plain.

Termite mounds are a common sight on the African veld. Some are shaped like narrow chimneys that sometimes tower over 20 feet. Others are large domes of earth that provide a favorite lookout post for predators such as lions. So writes a south African correspondent to the Awake magazine in the November 8 1993 issue, page 31.

He continues, Inside each mound are numerous passages and chambers,, which may be occupied by several million termite. Some termite tend their own fungal gardens and man- age to keep them well watered even during years of drought. During the 1930's when severe drought devastated parts of South Africa, a naturalist, Dr. Eugene Marais, discovered two columns of termite, one descending and the other ascending in a tunnel. The little creatures had burrowed to a depth of about 100 feet! They had reached a natural well. Thus Marais discovered how they managed to keep their fungal gardens moist through the drought. The artical continues, a typical termite mound, explains Michael Main in his book Kalahari, "is believed to be the most advanced nest built by any animal in the world....

All seek to achieve and maintain 100 per cent humidity and an ambient temperature between 29 degrees Celsius, [84 degrees fahrenheit] and 31 degrees Celsius [88 degrees fahrenheit], which suits both the fungus and the termite.... Every nest is, effectively, a perfect air- condition unit."

Now if you compare our size to the size of the mounds we construct and our heating and cooling systems, would you not agree that we lowly termite qualify to be called Ingenious Engineers- and that's not to mention the recycling of dead vegetation and thus dispose of much waste. Just remember, WE were created to help you, in caring for your responsibility in caring for our beautiful planet.

Now that you know why we are called Ingenious Engineers, what do you think? Are we Friends or Foes? I hope you said Friends, but before you answer let me tell you what a correspondent to the AWAKE magazine said in the May 22,1995 issue on page 18 said about this very matter.

The Termite Friend Or Foe?

While few people would deny that these insects are fascinat- ing, most still view them as pests--enemies! Dr. Richard Bag- ine, head of the Invertebrate Zoology Department of the Na- tional Museum of Kenya, told Awake!: "It is true that termites are seen by people as one of the most destructive insects.

But scientists see termites differently. In the wild, termites are useful members of the plant and animal community.

"First, they break down dead plant material into simple com- pounds. In this way, termites recycle nutrients that plants need. Second, they are an important food source. They are eaten by almost every kind of bird and by many mammals, reptiles, amphibians, and other insects. Many people in west- ern and northern Kenya also enjoy their sweet, rich taste; they are very rich in fats, and proteins.

Third, they help to make soil. Termites mix subsoil and topsoil when they build and repair their nests. They break down large pieces of dead plant material into smaller piece, forming humus. Moving through the soil, they make passages for air and water needed to plant roots. Thus termites improve soil texture, structure, and fertility."

Why, though, do termites invade human habitations? Says Dr. Bagine: "Actually, people have moved into the termites' habitats and removed most of the plant resoures used by ter- mites. Termites must eat to live, and they will usually feed on dead plants. When these are taken away from them, termites feed on man made structures, such as houses and granaries."

So although the termite may at times seem to be a pest, it surely is not our enemy. Indeed, it is a striking example of Jehovah's creative brilliance. (Psalm 148:10, 13; Romans 1:20) And in God's coming new world, as man learns to live in harmony with the animal world, he will no doubt come to see the tiny termite as a friend, not a foe.--Isaiah 65:25.

REFERENCES

Ebeling, Walter. "Wood-destroying Insects and Fungi." In *Urban Entomology*, 128-167.

IBSA. (Watchtower Bible & Tract Society of New York, Inc. International Bible Students Association). *Insight on the Scriptures*, vol. 2. Brooklyn, NY: IBSA.

Rutenberg, James. 1999. "Termites gnaw FDR drive pilings—'Gribbles' have major case of gobbles." *New York Daily News*, January 28, 1999.

Walker, Norman W. 1970. *Fresh Vegetables and Fruit Juices: What Is Missing in Your Body?* Prescott, AZ: Norwalk Press.

Walker, Norman W. 1995. *Colon Health: The Key to a Vibrant Life*. Prescott, AZ: Norwalk Press.

Wikipedia 2019. "Disodium Octaborate Tetrahydrate." Last modified 25 January 2019, at 19:05. https://en.wikipedia.org/wiki/Disodium octaborate tetrahydrate.

ADDITIONAL READING

Anderson, Richard. 1988. *Cleanse & Purify Thyself*; 2nd ed.

Batmanghelidj, F. 1977. *Your Body's Many Cries for Water*; 2nd ed. Falls Church, VA: Global Health Solutions, Inc.

Bennett, Gary W., John M. Owens, and Robert M. Corrigan. 1988. *Truman's Scientific Guide to Pest Control Operations*. Duluth, MN: Advanstar Communications.

Braggs, Paul C. and Patricia Braggs. *The Miracle of Fasting*. Santa Barbara, CA: Health Science.

Braggs, Paul C. and Patricia Braggs. *Apple Cider Vinegar*. Santa Barbara, CA: Health Science.

Braggs, Paul C. and Patricia Braggs. *Braggs Healthy Lifestyle*. Santa Barbara, CA: Health Science.

Braggs, Paul C. and Patricia Braggs. *Water—The Shocking Truth*. .Santa Barbara, CA: Health Science.

IBSA (Watchtower Bible & Tract Society of New York, Inc. International Bible Students Association). "Why Do Many Accept Evolution?" In *Life—How Did It Get Here? By Evolution or by Creation?* Brooklyn, NY: IBSA.

Jensen, Bernard. 1983. *Breathe Again Naturally: How to Deal with Catarrh, Bronchitis, Asthma & Manage Lung and Bronchial Problems Through a Natural Living & Eating Program*. Escondido, CA: Bernard Jensen Enterprises.

Jensen, Bernard and Sylvia Bell. 1981. *Tissue Cleansing Through Bowel Management*, 11th ed. Escondido, CA: Bernard Jensen Enterprises.

Lin, David J. 1993. *Free Radicals and Disease Prevention*. New Canaan, CO: Keats Publishing, Inc.

Walker, Norman W. 1978. *Become Younger.* Prescott, AZ: Norwalk Press.

Walker, Norman W. 1981. *Pure and Simple Natural Weight Control.* Prescott, AZ: Norwalk Press.

Walker, Norman W. 1995. *The Vegetarian Guide to Diet & Salad.* Prescott, AZ: Norwalk Press.

Keith, Velma J. and Monteen Gordon. *The How to Herb Book: Let's Remedy The Situation.*

Kofoid, Charles A., ed. 1934. *Termite and Termite Control,* 2nd ed. With S. F. Light, A. C. Homer, Merle Randall, W. B. Herms, and Earl E. Bowe. Berkeley, CA: University of California Press.

Penn State Extension. *Eastern Subterranean Termites.*

OSU (Ohio State University). "Bulletin Biology of Subterranean Termites in the Eastern United States, Bulletin #1209." OH: OSU.

The Encyclopedia Americana. 1829–1953 volume 26. s.v. "Termites, White Ants, or Duck Ants."

Watchtower (Watchtower Bible & Tract Society of New York, Inc.). 1991. *The Greatest Man Who Ever Lived.* Brooklyn, NY: Watchtower.

Watchtower (Watchtower Bible & Tract Society of New York, Inc.). 2005. "The Divine Name: It's Use and It's Meaning." In *What Does The Bible Really Teach?* 195. Brooklyn, NY: Watchtower.

UDA (Utah Department of Agriculture Division of Plant Industry). "Wood-Destroying Pest Control: Pesticide Application & Safety Training, Study Guide." UT: UDA.

ABOUT THE ILLUSTRATOR

Lorenz was born on the Navajo Reservation in Monument Valley, Utah. He resides on Holiday Mesa, which is just south of Oljato, located eight miles west of Goulding's Lodge.

As a child, his parents sent him to boarding school in Kayenta, Arizona, where he was taught English through reading and writing—an experience he did not relish at a young age. He preferred to look after his grandmother's sheep and roam the nearby hillsides where he could perfect his artistic talents.

Lorenz developed his self-taught talents of drawing, carving, and painting by observing his grandparents and parents participating in traditional Navajo ceremonies and creating sand paintings using natural pigments for various colors.

PICTURE GALLERY

Izzy's playground

Did you say termite tube

Izzy's Life Cycle

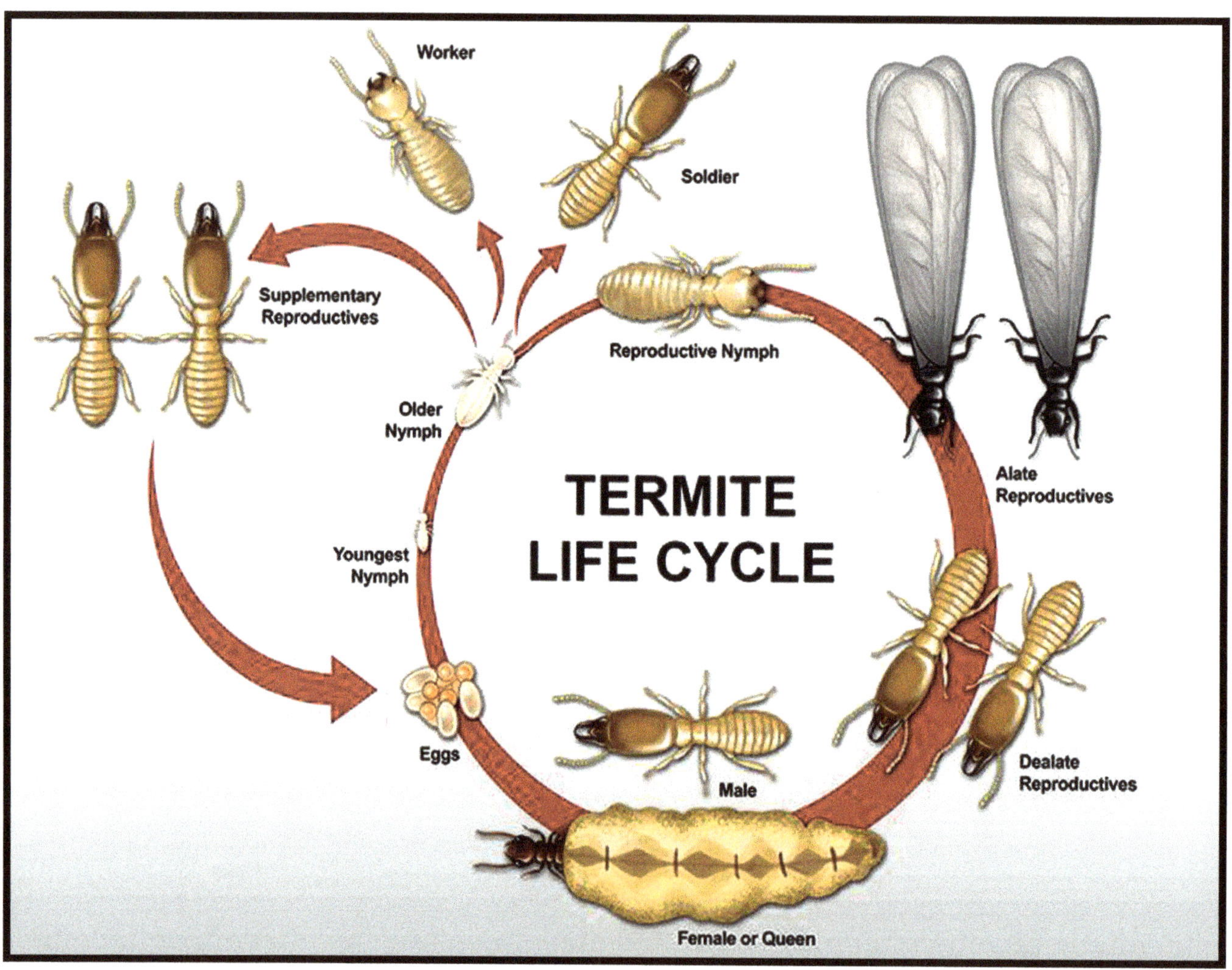

Izzy's Queen

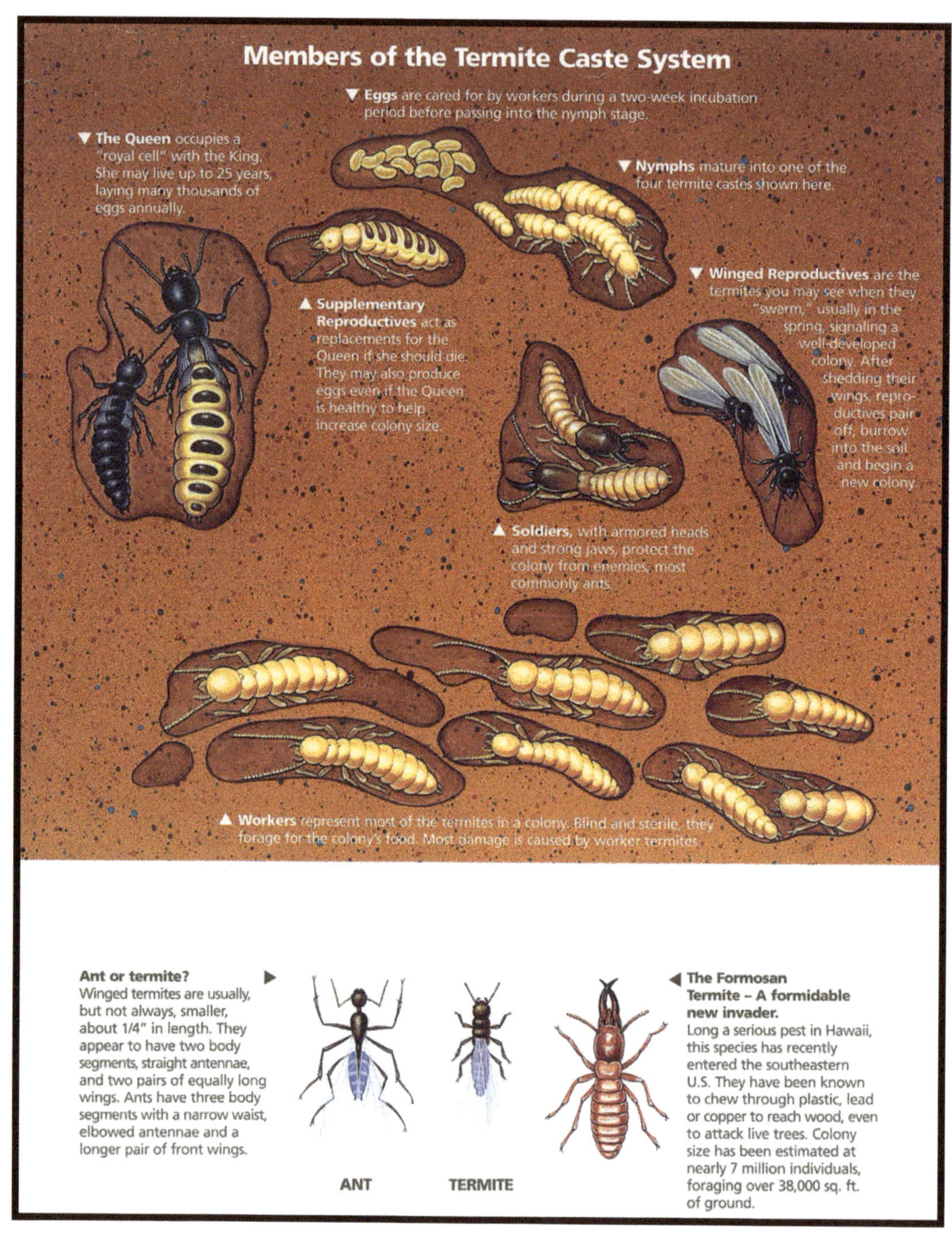

Members of the Termite Caste System
▼ Eggs are cared for by workers during a two-week incubation period before passing into the nymph stage.
▼ The Queen occupies a "royal cell" with the King. She may live up to 25 years, laying many thousands of eggs annually.
▼ Nymphs mature into one of the four termite castes shown here.
▲ Supplementary Reproductives act as replacements for the Queen if she should die. They may also produce eggs even if the Queen is healthy to help increase colony size.
▼ Winged Reproductives are the termites you may see when they "swarm," usually in the spring, signaling a well-developed colony. After shedding their wings, reproductives pair off, burrow into the soil and begin a new colony.
▲ Soldiers, with armored heads and strong jaws, protect the colony from enemies, most commonly ants.
▲ Workers represent most of the termites in a colony. Blind and sterile, they forage for the colony's food. Most damage is caused by worker termites.
Ant or termite? ▶
Winged termites are usually, but not always, smaller, about 1/4" in length. They appear to have two body segments, straight antennae, and two pairs of equally long wings. Ants have three body segments with a narrow waist, elbowed antennae and a longer pair of front wings.
ANT
TERMITE
◀ The Formosan Termite – A formidable new invader.
Long a serious pest in Hawaii, this species has recently entered the southeastern U.S. They have been known to chew through plastic, lead or copper to reach wood, even to attack live trees. Colony size has been estimated at nearly 7 million individuals, foraging over 38,000 sq. ft. of ground.

Izzy's Workers

Izzy's Soldiers

Izzy's Swarmers

See if you could pick out the soldiers

NOTES

With my Daughter Nicole
At My side science she was in diapers

With Dr. Austin M. Frishman

The grand pa of our industry
In the early 70'S he suggested,
"to be a good operator, you have to want to"

Sylvia Kenmuir
The Queen Entomologist
Board Certified Entomologist

With Nicole, Dr. Frishman and I
Dr. Frishman last presentation before retirement

My bride of 53 years and everlasting to go on the planet Earth. Rev. 21:3 & 4
My walking field manual and insect identification guide

Illustrated by

Lorenz and Barbara Holiday

Lorenz was born December 18, 1966, on the Navajo Reservation in Rock Door Canyon, Utah. He attended boarding school in Kayenta, Arizona.

He developed his self-taught talents of drawing and painting by watching his grandparents create sand paintings using natural pigments.

NOTES

ABOUT THE AUTHOR

James R. Melendez

There is nothing special about him, other than the fact that he is very dedicated to Bible truth. Because of that, it permeates every aspect of his life. The termite industry is no exception.

His formal training was in the field of music (Berklee College of Music, US Navy school of music, and Navy band), which laid the groundwork for logical training, a must in termite entomology. He studied pest control through Purdue University

Home Correspondence and the University of California, Berkeley and worked in close associations with the Extension Service of Arizona State University, Northern Arizona State University, Utah State University, NPMA, and local state compliance agencies.

His emphasis has always been to work in ways that are safer, more efficient, and most of all more environmentally friendly to all.

For questions or information on any of the substances mentioned in this volume, please feel free to contact me at:

James R. Melendez
P. O. Box 95
Monticello, UT 84535
435-587-2165
888-673-7645
480-861-6093 mobile
Feedyoursoil.com

www.ingramcontent.com/pod-product-compliance
Ingram Content Group UK Ltd.
Pitfield, Milton Keynes, MK11 3LW, UK
UKHW061951290726
14090UKWH00021B/1176

9 798886 406535